I0488116

Determination of Steroid Hormones and Related Compounds in Filtered and Unfiltered Water by Solid-Phase Extraction, Derivatization, and Gas Chromatography with Tandem Mass Spectrometry

By William T. Foreman, James L. Gray, Rhiannon C. ReVello, Chris E. Lindley, Scott A. Losche, and Larry B. Barber

Techniques and Methods 5–B9

U.S. Department of the Interior
U.S. Geological Survey

U.S. Department of the Interior
KEN SALAZAR, Secretary

U.S. Geological Survey
Marcia K. McNutt, Director

U.S. Geological Survey, Reston, Virginia: 2012

For more information on the USGS—the Federal source for science about the Earth, its natural and living resources, natural hazards, and the environment, visit http://www.usgs.gov or call 1–888–ASK–USGS.
For an overview of USGS information products, including maps, imagery, and publications,
visit http://www.usgs.gov/pubprod

To order this and other USGS information products, visit http://store.usgs.gov

Suggested citation:
Foreman, W.T., Gray, J.L., ReVello, R.C., Lindley, C.E., Losche, S.A., and Barber, L.B., 2012, Determination of steroid hormones and related compounds in filtered and unfiltered water by solid-phase extraction, derivatization, and gas chromatography with tandem mass spectrometry: U.S. Geological Survey Techniques and Methods, book 5, chap. B9, 118 p.

Electronic versions of the preparatory and analytical standard operating procedures for the methods are available upon request to *LabHelp@usgs.gov.*

Contents

iv

Figures

Tables

Conversion Factors and Abbreviations

Pound/Inch to SI

Multiply	By	To obtain
Length		
centimeter (cm)	0.3937	inch (in.)
micrometer (µm)	3.937×10^{-5}	inch (in.)
millimeter (mm)	0.03937	inch (in.)
meter (m)	3.281	foot (ft)
Volume		
liter (L)	33.82	ounce, fluid (fl. oz)
liter (L)	2.113	pint (pt)
liter (L)	1.057	quart (qt)
liter (L)	0.2642	gallon (gal)
liter (L)	61.02	cubic inch (in^3)
milliliter (mL)	0.0338	ounce, fluid (fl. oz)
microliter (µL)	0.0338×10^{-3}	ounce, fluid (fl. oz)
Mass		
gram (g)	0.03527	ounce, avoirdupois (oz)
microgram (µg)	3.53×10^{-8}	ounce, avoirdupois (oz)
milligram (mg)	3.53×10^{-5}	ounce
nanogram (ng)	3.53×10^{-11}	ounce
picogram (pg)	3.53×10^{-14}	ounce
Pressure		
kilopascal (kPa)	0.009869	atmosphere, standard (atm)
kilopascal (kPa)	0.01	bar
kilopascal (kPa)	0.2961	inch of mercury at 60°F (in Hg)
kilopascal (kPa)	0.1450	pound per square inch (lb/in^2)
kilopascal (kPa)	1,000	pascal (Pa)
pascal (Pa)	1,000	millipascal (mPa)
Concentration, in water		
nanograms per liter (ng/L)	1	parts per trillion
milligrams per liter (mg/L)	1	parts per million

Temperature in degrees Celsius (°C) may be converted to degrees Fahrenheit (°F) as follows:

°F=(1.8×°C)+32

Specific conductance is given in microsiemens per centimeter at 25 degrees Celsius (µS/cm at 25°C).

Concentrations of chemical constituents in water in this report are given in nanograms per liter (ng/L).

Abbreviated Water-Quality Units

μg/μL   microgram per microliter

mg/L   milligram per liter

ng/L   nanogram per liter

ng/μL   nanogram per microliter

pg/μL   picogram per microliter

Acronyms, Abbreviations, and Symbols

α	alpha
β	beta
BPA	bisphenol A
^{13}C	carbon-13 isotope
C_{18}	octadecyl
C_{18} disk	glass-fiber filter disk embedded with octadecyl surface-modified silica
cat. no.	catalog number
CCV	continuing calibration verification
C=O	carbonyl functional group; specifically ketone group for the method compounds
C–OH	alcohol functional group
d	deuterium
DCM	dichloromethane
DOC	dissolved organic carbon
E	estimated result-level remark code used in the National Water Information System
FRLMS	field-requested laboratory matrix-spike sample
GC	gas chromatography
GC/MS/MS	gas chromatography/tandem mass spectrometry
GFF	glass-fiber filter
h	hour
HDPE	high-density polyethylene
i.d.	inner diameter
IDE	interlaboratory detection estimate (also intralaboratory detection estimate in this report)
IDQ	isotope-dilution quantification
IDS	isotope-dilution standard

IIS	injection internal standard
IRL	interim reporting level
Lc	Currie's critical level
LRB	laboratory reagent blank or laboratory reagent-water blank in this report
LRL	laboratory reporting level
LRS	laboratory reagent spike or laboratory reagent-water spike in this report
LS	laboratory schedule
M^+	molecular ion
MDL	method detection limit
DUP	laboratory matrix duplicate sample
min	minute
MRL	minimum reporting level
MRM	multiple-reaction monitoring
MS/MS	tandem-quadrupole mass spectrometry
MSPK	laboratory matrix-spike sample
MSTFA	N-methyl-N-trimethylsilyl trifluoroacetamide
m/z	mass-to-charge ratio
N_2	nitrogen gas
NWIS	National Water Information System
NWQL	National Water Quality Laboratory
no.	number
o.d.	outer diameter
QA	quality assurance
QA/QC	quality assurance/quality control
QC	quality control
RF_σ	relative F-pseudosigma
RPD	relative percent difference (modulus; absolute value of)
PC	percentage change
RSD	relative standard deviation
RWS	reagent-water spike
SOP	standard operating procedure
SPE	solid-phase extraction
SSE	stainless-steel extractor
™	trademark
USEPA	U.S. Environmental Protection Agency

USGS	U.S. Geological Survey
WWTP	wastewater-treatment plant
$\equiv$	identical with, congruent
>	greater than
$\geq$	greater than or equal to
<	less than
$\leq$	less than or equal to
$\pm$	plus or minus
$\times$	times (multiplication)
/	per; divided by
®	registered trademark

Determination of Steroid Hormones and Related Compounds in Filtered and Unfiltered Water by Solid-Phase Extraction, Derivatization, and Gas Chromatography with Tandem Mass Spectrometry

By William T. Foreman, James L. Gray, Rhiannon C. ReVello, Chris E. Lindley, Scott A. Losche, and Larry B. Barber

Abstract

A new analytical method has been developed and implemented at the U.S. Geological Survey National Water Quality Laboratory that determines a suite of 20 steroid hormones and related compounds in filtered water (using laboratory schedule 2434) and in unfiltered water (using laboratory schedule 4434). This report documents the procedures and initial performance data for the method and provides guidance on application of the method and considerations of data quality in relation to data interpretation. The analytical method determines 6 natural and 3 synthetic estrogen compounds, 6 natural androgens, 1 natural and 1 synthetic progestin compound, and 2 sterols: cholesterol and 3-*beta*-coprostanol. These two sterols have limited biological activity but typically are abundant in wastewater effluents and serve as useful tracers. Bisphenol A, an industrial chemical used primarily to produce polycarbonate plastic and epoxy resins and that has been shown to have estrogenic activity, also is determined by the method.

A technique referred to as isotope-dilution quantification is used to improve quantitative accuracy by accounting for sample-specific procedural losses in the determined analyte concentration. Briefly, deuterium- or carbon-13-labeled isotope-dilution standards (IDSs), all of which are direct or chemically similar isotopic analogs of the method analytes, are added to all environmental and quality-control and quality-assurance samples before extraction. Method analytes and IDS compounds are isolated from filtered or unfiltered water by solid-phase extraction onto an octadecylsilyl disk, overlain with a graded glass-fiber filter to facilitate extraction of unfiltered sample matrices. The disks are eluted with methanol, and the extract is evaporated to dryness, reconstituted in solvent, passed through a Florisil solid-phase extraction column to remove polar organic interferences, and again evaporated to dryness in a reaction vial. The method compounds are reacted with activated *N*-methyl-*N*-trimethylsilyl trifluoroacetamide at 65 degrees Celsius for 1 hour to form trimethylsilyl or trimethylsilyl-enol ether derivatives that are more amenable to gas chromatographic separation than the underivatized compounds. Analysis is carried out by gas chromatography with tandem mass spectrometry using calibration standards that are derivatized concurrently with the sample extracts.

Analyte concentrations are quantified relative to specific IDS compounds in the sample, which directly compensate for procedural losses (incomplete recovery) in the determined and reported analyte concentrations. Thus, reported analyte concentrations (or analyte recoveries for spiked samples) are corrected based on recovery of the corresponding IDS compound during the quantification process. Recovery for each IDS compound is reported for each sample and represents an absolute recovery in a manner comparable to surrogate recoveries for other organic methods used by the National Water Quality Laboratory. Thus, IDS recoveries provide a useful tool for evaluating sample-specific analytical performance from an absolute mass recovery standpoint. IDS absolute recovery will differ and typically be lower than the corresponding analyte's method recovery in spiked samples. However, additional correction of reported analyte concentrations is unnecessary and inappropriate because the analyte concentration (or recovery) already is compensated for by the isotope-dilution quantification procedure.

Method analytes were spiked at 10 and 100 nanograms per liter (ng/L) for most analytes (10 times greater spike levels were used for bisphenol A and 100 times greater spike levels were used for 3-*beta*-coprostanol and cholesterol) into the following validation-sample matrices: reagent water, wastewater-affected surface water, a secondary-treated wastewater effluent, and a primary (no biological treatment) wastewater effluent. Overall method recovery for all analytes in these matrices averaged 100 percent, with overall relative standard deviation of 28 percent. Mean recoveries of the 20 individual analytes for spiked reagent-water samples prepared along with field samples and analyzed in 2009–2010 ranged from 84–104 percent, with relative standard deviations of 6–36 percent. Concentrations for two analytes, equilin and progesterone, are reported as estimated because these analytes had excessive bias or variability, or both. Additional database coding is applied to other reported analyte data as needed, based on sample-specific IDS recovery performance.

Detection levels were derived statistically by fortifying reagent water at six different levels (0.1 to 4 ng/L) and range from about 0.4 to 4 ng/L for 16 analytes. Interim reporting levels applied to analytes in this report range from 0.8 to 8 ng/L. Bisphenol A and the sterols (cholesterol and 3-*beta*-coprostanol) were consistently detected in laboratory and field blanks. The minimum reporting levels were set at 100 ng/L for bisphenol A and at 200 ng/L for the two sterols to prevent any bias associated with the presence of these compounds in the blanks. A minimum reporting level of 2 ng/L was set for 11-ketotestosterone to minimize false positive risk from an interfering siloxane compound emanating as chromatographic-column bleed, from vial septum material, or from other sources at no more than 1 ng/L.

Introduction

Over the last 15 years, an increasing number of scientific investigations have documented the potential of estrogenic hormones to affect the endocrine systems of exposed organisms at extremely low doses; in some cases at less than 1 nanogram per liter (ng/L) (Routledge and others, 1998; Lange and others, 2001; McGee and others, 2009; and reviews by Mills and Chichester, 2005; and Caldwell and others, 2008). The primary pathways by which steroid hormones are introduced to the environment include discharge of municipal and industrial wastewater and runoff from agricultural operations, although a large variety of anthropogenic sources has been considered (Sumpter and Johnson, 2005). These compounds can occur in the environment at concentrations exceeding published lowest-observable effects concentrations, especially in treated wastewater effluents and surface waters that receive discharge from wastewater-treatment plants (WWTPs) (for example, see Ternes and others, 1999; Huang and Sedlak, 2001).

Furthermore, collaborative studies by the U.S. Geological Survey (USGS) (Vajda and others, 2008, 2011) and studies by others (for example, Jobling and others, 1998; Tilton and others, 2002) have shown that fish living downstream from some WWTP discharges have abnormal development of sex organs, and that exposure to natural and synthetic estrogens is likely to play a role in the induction of such abnormalities. These effects, known as endocrine disruption, can be manifested in several different ways including inappropriate expression of vitellogenin (an egg yolk protein) by males or juveniles, demasculinization of secondary sex characteristics, suppression of gonadal development, suppression of sperm development, and the formation of intersex gonadal tissue, which occurs when both male and female reproductive germ tissue are present in the same individual.

The evidence of biological activity of steroids at environmental concentrations and subsequent deleterious effects on aquatic organisms is strongest for the estrogens, particularly the principal human estrogen, 17-*beta*-estradiol, its metabolite estrone, and the synthetic pharmaceutical 17-*alpha*-ethynylestradiol (Wise and others, 2011). For example, Kidd and others (2007) observed the total collapse of a fathead minnow population in a lake exposed to 6 ng/L of 17-*alpha*-ethynylestradiol. Mixtures of estrogens might even act additively or synergistically (Thorpe and others, 2006; Rajapakse and others, 2004), and consideration of all compounds possible that have known activity is ideal. Although there is less direct evidence of activity in the environment, androgens and progestins can induce biological effects by similar nuclear receptor-mediated modes-of-action and might exert similar effects at low concentrations (Ankley and others, 2003; Zeilenger and others, 2009).

As a result of these observations, there has been considerable interest within the USGS to provide the analytical capability to measure these compounds at environmentally relevant concentrations, in part, to (1) further understand their presence and distribution in the environment, (2) examine their role in inducing deleterious effects on wildlife, and (3) assess the efficacy of their removal from waste streams using various treatment technologies. To meet this need, the USGS National Water Quality Laboratory (NWQL) has developed a method to analyze for a suite of 20 target compounds (referred to as "analytes" throughout this report) in filtered and unfiltered water. The method is based on solid-phase extraction (SPE) of a water sample using octadecylsilyl (C_{18}) silica sorbent, removal of some coextracted compounds using Florisil SPE, chemical derivatization of method compounds by silylation of active functional groups, and analysis by gas chromatography with tandem-quadrupole mass spectrometry (GC/MS/MS).

Method analytes include 6 natural and 3 synthetic estrogens, 6 natural androgens, 1 natural and 1 synthetic progestin, 2 sterols, and the industrial chemical bisphenol A (BPA) that is known to have estrogenic activity (Vandenberg and others, 2009) (table 1; figure 1). The determined analytes include the seven hormones (17β-estradiol, 17α-ethynylestradiol, estriol, estrone, equilin, 4-androstene-3,17-dione, and testosterone) recently proposed under the revisions to the Unregulated Contaminant Monitoring Regulation (UCMR 3) for public water systems (U.S. Environmental Protection Agency, 2011).

Eleven of the 13 natural hormones included in the method are excreted by humans in free or conjugated forms in urine and feces (Liu and others, 2009a). Two others, equilin and equilenin, are equine hormones that are isolated and administered pharmaceutically during estrogen-replacement therapy (Belchetz, 1994). The four synthetic hormones have human pharmaceutical uses, although diethylstilbestrol use is limited due to its undesirable teratogenic side effects (Mittendorf, 1995). Only the strongly estrogenic (endocrine disrupting) *trans*-isomer of diethylstilbestrol is determined by the method. Note: reference to diethylstilbestrol (or

Table 1. Analyte names, Chemical Abstract Service (CAS) Registry Numbers, class, source or use, and codes used in the analytical method for determination of steroid hormones and related compounds.

[NWIS, National Water Information System; NWQL, National Water Quality Laboratory; --, not applicable]

Analyte name[a]	Short name or abbreviation[b]	CAS Registry Number[c]	Class	Source, use, and other comment	NWIS parameter and method code NWQL schedule 2434	4434
11-Ketotestosterone	Ketotestosterone	564–35–2	Natural androgen	Testosterone metabolite	64507GM004	64527GM005
17-*alpha*-Estradiol	17α-Estradiol; α-E2	57–91–0	Natural estrogen	Low human occurrence, common in some other species	64508GM004	64528GM005
17-*alpha*-Ethynylestradiol	Ethynylestradiol; EE2	57–63–6	Synthetic estrogen	Used in oral contraceptives	64509GM004	64529GM005
17-*beta*-Estradiol	17β-Estradiol; β-E2; E2	50–28–2	Natural estrogen	Principal estrogen in humans	64510GM004	64530GM005
3-*beta*-Coprostanol	3β-Coprostanol	360–68–9	Natural sterol	Carnivore fecal indicator; useful sewage tracer	64512GM004	64532GM005
4-Androstene-3,17-dione	Androstenedione; ADSD	63–05–8	Natural androgen	Testosterone precursor; illicit steroid	64513GM004	64533GM005
Bisphenol A	BPA	80–05–7	--	Monomer used to make polycarbonate plastics and epoxy resins	67304GM004	67305GM005
Cholesterol	--	57–88–5	Natural sterol	Ubiquitous; produced by animals and plants	64514GM004	64534GM005
cis-Androsterone	Androsterone	53–41–8	Natural androgen	Testosterone metabolite; used in deer repellant	64515GM004	64535GM005
Dihydrotestosterone	DHT	521–18–6	Natural androgen	Testosterone metabolite	64524GM004	64544GM005
Epitestosterone	EPI	481–30–1	Natural androgen	Testosterone isomer; human androgen	64517GM004	64537GM005
Equilenin	--	517–09–9	Natural estrogen	Equine estrogen; used in hormone replacement therapy	64518GM004	64538GM005
Equilin	--	474–86–2	Natural estrogen	Equine estrogen; used in hormone replacement therapy	64519GM004	64539GM005
Estriol	E3	50–27–1	Natural estrogen	Metabolite of 17β-estradiol	64520GM004	64540GM005
Estrone	E1	53–16–7	Natural estrogen	Metabolite of 17β-estradiol	64521GM004	64541GM005
Mestranol	--	72–33–3	Synthetic estrogen	Used in oral contraceptives; metabolized to ethynylestradiol prior to excretion	64522GM004	64542GM005
Norethindrone	--	68–22–4	Synthetic progestin	Used in oral contraceptives	64511GM004	64531GM005
Progesterone	--	57–83–0	Natural progestin	Principal human progestational hormone	64523GM004	64543GM005
Testosterone	--	58–22–0	Natural androgen	Principal human androgen	64525GM004	64545GM005
trans-Diethylstilbestrol[d]	Diethylstilbestrol; DES	56–53–1	Synthetic estrogen	Pharmaceutical	64516GM004	64536GM005

[a]The reporting unit in NWIS for analyte concentration is nanograms per liter.

[b]The short name is an alternative used at the NWQL and does not necessarily match the NWIS short name. Some steroid abbreviations are those frequently used in the literature.

[c]This report contains CAS Registry Numbers®, which is a Registered Trademark of the American Chemical Society. CAS recommends the verification of the Chemical Abstract Service Registry Numbers through CAS Client Services.

[d]*trans*-Diethylstilbestrol is the only isomeric form of diethylstilbestrol being determined by the method.

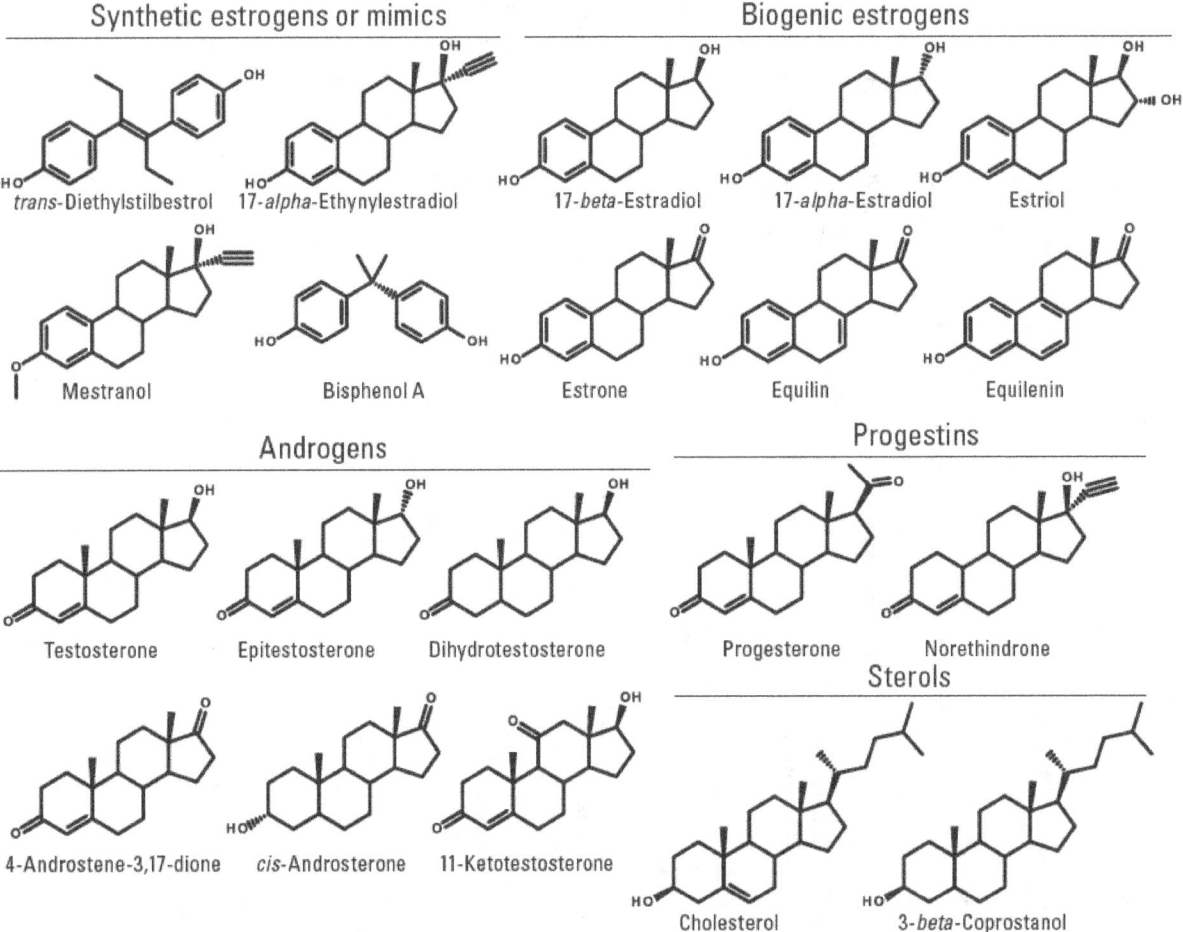

Figure 1. Structures of method analytes.

its isotope-dilution standard diethylstilbestrol-d_8) in this report is for the *trans*-isomer; *cis*-diethylstilbestrol (or *cis*-diethylstilbestrol-d_8) is not determined.

The two sterols, cholesterol and 3β-coprostanol, typically are present at high concentrations in waters receiving substantial WWTP discharges, runoff from fields with applied biosolids, and runoff from animal feeding operations, and are included mainly as potential indicators of contamination (Ayebo and others, 2006) and not as primary target analytes for this method, which was designed specifically to determine substantially lower concentrations of the steroid hormones. The method determines the free (non-conjugated) forms of the method analytes. Most of the glucoronide conjugates of the steroids are converted to the free form during WWTP processes, whereas sulfate conjugates are partially (35–88 percent) deconjugated (Kumar and others, 2011).

Although discharge of WWTP effluents to surface waters is a major source of these hormones to the environment, these effluents are by no means the only potential source. Considerable effort has been made to ensure that this method is robust and sensitive for a variety of water matrices,

including those that may contain substantial amounts of organic matter and other potential interferences. Both filtered and unfiltered water samples have been analyzed, with the majority of sample submissions being unfiltered. Water-sample types analyzed in the course of method validation and custom-sample analysis include: groundwater; runoff from agricultural land amended with biosolids from wastewater treatment; WWTP effluents, influents, and internal-recycle streams; both treated and untreated combined-sewer overflows; various surface waters including those with high concentrations of dissolved organic carbon (DOC) and suspended sediment; and streams affected by runoff from animal-feeding operations.

The method incorporates several techniques not used in previous organic-chemistry methods at the NWQL that contribute to enhanced specificity, selectivity, and reliability, especially in difficult matrices. This report documents these techniques, describes the advantages gained through their implementation, and provides some guidance to customers on how best to use and interpret the results produced by the method.

First, samples undergo chemical derivatization with activated N-methyl-N-trimethylsilyl trifluoroacetamide (MSTFA) before GC/MS/MS analysis. Derivatization makes the method compounds less polar and more volatile and, thus, readily amenable to gas chromatography (GC) separation. These higher molecular weight trimethylsilyl or trimethylsilyl-enol ether derivatives also produce characteristic mass spectrometric ions of higher mass-to-charge ratio (m/z) that typically make identification and quantitation less susceptible to interference.

Second, the application of tandem-quadrupole mass spectrometry (MS/MS) in comparison to single-quadrupole mass spectrometry dramatically improves the specificity of the analysis and decreases the likelihood of false positives. The NWQL method described in this report evolved from a method developed by Barber and others (2000) of the USGS National Research Program that used single-quadrupole mass spectrometry, and substantial improvements in selectivity and achievable detection levels were obtained using MS/MS. Indeed, over the last decade, many methods for determining steroid hormones by either GC or liquid chromatography used either MS/MS (for example, Kelly, 2000; Huang and Sedlak, 2001; Ternes and others, 2002; Fine and others, 2003; Ingrand and others, 2003; Carpinteiro and others, 2004; Rodriquez-Mozaz and others, 2004; Tolgyesi and others, 2010; U.S. Environmental Protection Agency, 2010a) or high-resolution mass spectrometry (for example, Hohenblum and others, 2004; U.S. Environmental Protection Agency, 2007a) for their high qualitative identification and low detection-level capabilities.

Finally, analyte concentrations are determined by using isotope-dilution quantification (IDQ), a procedure also applied in some U.S. Environmental Protection Agency (USEPA) methods—for example, USEPA method 8290A for polychlorinated dioxins and furans (U.S. Environmental Protection Agency, 2007b) and USEPA methods 1698 and 539 for selected steroid hormones (U.S. Environmental Protection Agency, 2007a, 2010a). For the NWQL method, isotope-dilution standard (IDS) compounds are added to the samples before extraction. These IDS compounds are deuterium- or carbon-13 (^{13}C)-labeled compounds (table 2; figure 2), 10 of which are direct isotopic analogs of the method analytes. The remaining method analytes also are quantified using isotope dilution by using one of the other four IDS compounds that have similar related chemical functionality but are not direct isotopic analogs of the analyte. (Note. cis-androsterone-2,2,3,4,4-d_5, estriol-2,4,16,17-d_4, and progesterone-2,3,4-$^{13}C_3$, listed in table 2 and figure 2 were not available for initial method validation. These three isotopes have been added to the method as exact IDS compounds for the quantification of cis-androsterone, estriol, and progesterone, respectively.)

Chemical behavior during sample preparation is considered to be nearly identical between a given method analyte and its corresponding IDS. Instead of calculating analyte concentrations based on quantitation relative to a traditional injection internal standard (IIS), as is used for many other NWQL organic methods, each result is determined relative to a corresponding IDS. This has the benefit of automatically correcting the analyte concentration for procedural losses due to many factors, including incomplete extraction, partial extract spills, low derivatization yield, matrix interferences, instrumental signal suppression or enhancement, or other mechanisms. That is, any biases resulting from incomplete recovery during sample preparation are corrected for before reporting analyte data by normalizing the analyte concentration to the IDS recovery.

The IDQ approach in this method represents a substantial difference in how the analyte concentration is determined compared to other organic methods provided by the NWQL. For existing NWQL organic methods, several surrogate compounds fortified into every sample are reported in percent recoveries as an indicator of overall method performance, but analyte concentrations are not corrected relative to the surrogate recoveries because the surrogates do not chemically emulate all of the method analytes. In this method for steroid hormones, the IDS compounds are functionally similar to traditional surrogates: they are added before any sample processing, carried through sample preparation, and their recoveries are used as a direct, sample-specific measure of method performance. The IDS recoveries for a sample also are reported (in percent recovery) to the USGS National Water Information System (NWIS) database; the public-accessible version of this database, NWISweb, is available at *http://waterdata.usgs.gov/nwis* (accessed March 2012). However, IDS recoveries are a measure of sample-specific absolute method recovery of the IDS (and corresponding analytes) and, as such, will differ from and might be substantially lower than the corresponding analyte's method recovery in spiked samples. However, additional correction of reported analyte concentrations is unnecessary and inappropriate because the analyte concentration or recovery already is compensated for by the IDQ procedure.

Purpose and Scope

This report describes the analytical method for the determination of selected steroid hormones and other compounds in filtered water (USGS method number O–2434–12) and unfiltered water (USGS method number O–4434–12). The report includes a brief outline of procedures to use for field collection, processing, and shipment of the water samples to the NWQL. The report also summarizes a set of validation studies, including spike recovery experiments in reagent water and three field matrices that are anticipated to be submitted frequently for analysis: a surface-water sample (Rapid Creek) collected immediately downstream from a WWTP in Rapid City, S. Dak.; one secondary WWTP effluent sample; and one primary WWTP effluent sample. The primary effluent sample was collected from the headworks of a WWTP after particle settling but before any biological treatment (Phillips and others, 2010), so this sample was substantially higher

Table 2. Isotope-dilution standard (IDS) and surrogate names, Chemical Abstract Service (CAS) Registry Numbers, analyte quantified relative to the IDS, and codes used in the analytical method for determination of steroid hormones and related compounds.

[NA, not available; NWIS, National Water Information System; NWQL, National Water Quality Laboratory; --, not applicable]

Compound long name in NWIS	Short name or abbreviation[a]	CAS Registry Number	Analyte quantified using IDS	NWIS parameter and method code NWQL schedule	
				2434	4434
Isotope-dilution standards[b]					
17-*alpha*-Ethynylestradiol-2,4,16,16-d_4	17-*alpha*- or 17α-Ethynylestradiol-d_4	350820–06–3	17α-Ethynylestradiol	90813GM004	90827GM005
17-*beta*-Estradiol-13,14,15,16,17,18-$^{13}C_6$	Estradiol-$^{13}C_6$	NA	17α- and 17β-Estradiol; equilenin	90777GM004	90780GM005
Bisphenol A-d_{16}	Bisphenol A-d_{16}	96210–87–6	Bisphenol A	67308GM004	67309GM005
Cholesterol-25,26,26,26,27,27,27-d_7	Cholesterol-d_7	83199–47–7	Cholesterol; 3β-coprostanol	90778GM004	90781GM005
cis-Androsterone-2,2,3,4,4-d_5[c]	*cis*-Androsterone-d_5[c]	NA	*cis*-Androsterone[c]	90816GM004	90509GM005
Estriol-2,4,16,17-d_4[d]	Estriol-d_4[d]	NA	Estriol[d]	91615GM004	91616GM005
Estrone-13,14,15,16,17,18-$^{13}C_6$	Estrone-$^{13}C_6$	NA	Estrone and equilin	90820GM004	90835GM005
Mestranol-2,4,16,16-d_4	Mestranol-d_4	NA	Mestranol	90821GM004	90836GM005
Nandrolone-16,16,17-d_3	Nandrolone-d_3	120813–22–1	4-Androstene-3,17-dione; dihydrotestosterone; testosterone; 11-ketotestosterone; epitestosterone; norethindrone; *cis*-androsterone[c]	91679GM004	91683GM005
Progesterone-2,3,4-$^{13}C_3$[e]	Progesterone-$^{13}C_3$[e]	327048–87–3	Progesterone[e]	90510GM004	90511GM005
trans-Diethyl-1,1,1',1'-d_4-stilbestrol-3,3',5,5'-d_4[f]	Diethylstilbestrol-d_4[f], DES-d_8	91318–10–4	*trans*-Diethylstilbestrol[f]	90817GM004	90832GM005
Surrogates[b]					
16-Epiestriol-2,4-d_2[d]	16-Epiestriol-d_2[d]	366495–94–5	Estriol[d]	91676GM004[g]	91680GM005[g]
Medroxyprogesterone-d_3[e]	Medroxyprogesterone-d_3[e]	162462–69–3	Progesterone[e]	91678GM004[g]	91682GM005[g]
Other reported property					
Sample volume, in milliliters	--	--	--	91118GM004	91119GM005

[a]The short name(s) is an alternative used in this report or by the NWQL and does not necessarily match the NWIS short name. The preceding stereoisomer indicator 17-*alpha*- or 17-*beta* also is omitted in some IDS short names used in this report.

[b]The reporting unit in NWIS for IDS and surrogate compounds is percent recovery.

[c]*cis*-Androsterone-2,2,3,4,4-d_5 was implemented as the IDS for quantifying *cis*-androsterone on October 1, 2011. For the validation data summarized in this report, nandrolone-d_3 was used as the IDS for quantifying *cis*-androsterone (see section 10.7).

[d]Estriol-2,4,16,17-d_4 was implemented as the IDS for quantifying estriol on March 17, 2011, at which time the stereoisomer 16-epiestriol-d_2 was changed from use as an IDS to a surrogate compound. For the validation data summarized in this report, 16-epiestriol-d_2 was used as the IDS for quantifying estriol (see section 10.7).

[e]Progesterone-2,3,4-$^{13}C_3$ was implemented as the IDS for quantifying progesterone on March 1, 2012, at which time medroxyprogesterone-d_3 was changed from use as an IDS to a surrogate compound. For the validation data summarized in this report, medroxyprogesterone-d_3 was used as the IDS for quantifying progesterone (see section 10.7).

[f]Only the *trans*-diethylstilbestrol-d_8 isomer is determined by the method. All references to diethylstilbestrol-d_8 in this report are indicating the *trans*-isomer.

[g]This NWIS parameter code also was used when the compound was previously classified as an IDS compound.

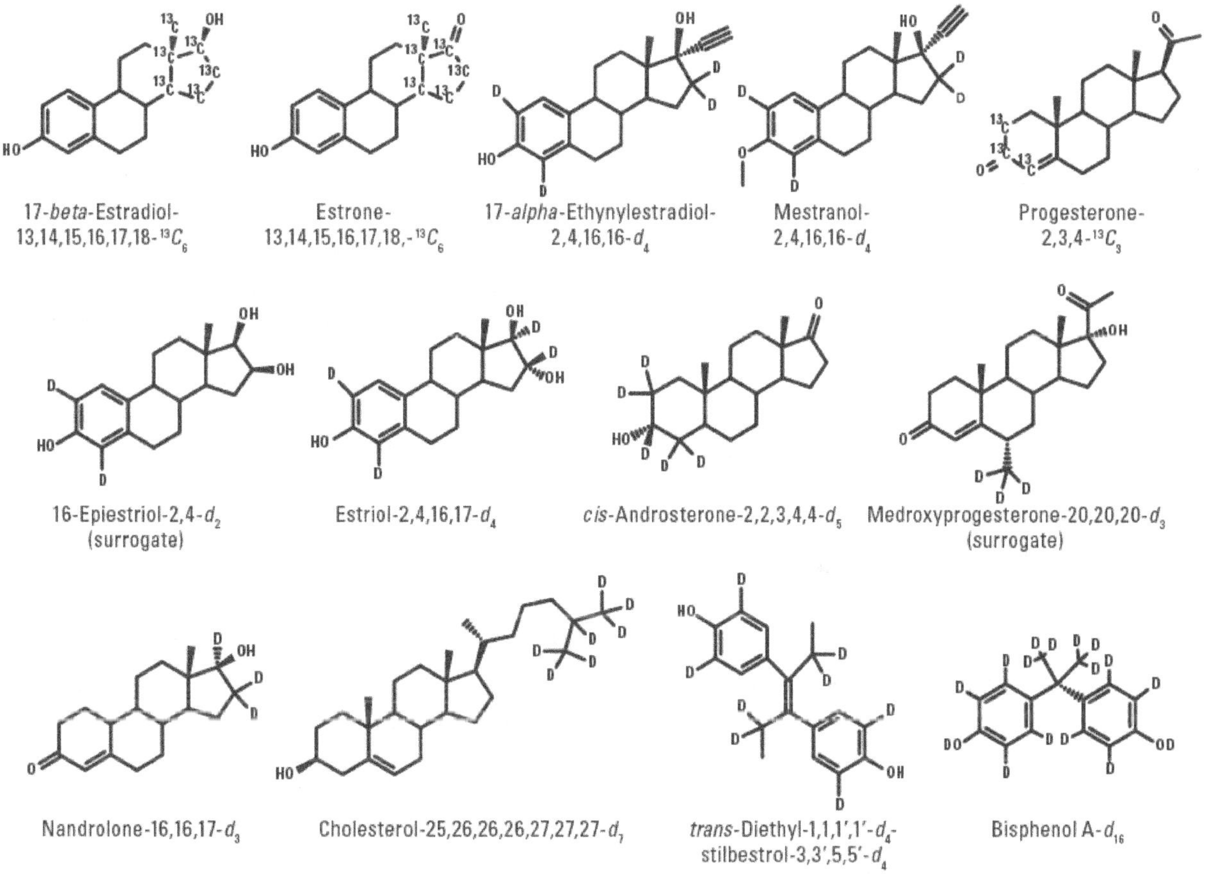

Figure 2. Structures of isotope-dilution standard and surrogate compounds.

17-*beta*-Estradiol-
13,14,15,16,17,18-$^{13}C_6$

Estrone-
13,14,15,16,17,18,-$^{13}C_6$

17-*alpha*-Ethynylestradiol-
2,4,16,16-d_4

Mestranol-
2,4,16,16-d_4

Progesterone-
2,3,4-$^{13}C_3$

16-Epiestriol-2,4-d_2
(surrogate)

Estriol-2,4,16,17-d_4

cis-Androsterone-2,2,3,4,4-d_5

Medroxyprogesterone-20,20,20-d_3
(surrogate)

Nandrolone-16,16,17-d_3

Cholesterol-25,26,26,26,27,27,27-d_7

trans-Diethyl-1,1,1',1'-d_4-
stilbestrol-3,3',5,5'-d_4

Bisphenol A-d_{16}

in particulate matter, DOC, and concentrations of selected analytes than the other test matrices.

This report also presents the results of a multi-concentration detection level determination, results of sample and extract holding-time experiments, and results from custom analysis of samples in a variety of aqueous matrices. Method data-reporting procedures are addressed, including those applied to laboratory and field-based quality-control (QC) and quality-assurance (QA) samples.

Note. all IDS data presented in this report are absolute IDS recoveries obtained for the isotope. The IDS absolute recovery is the determined mass of the IDS relative to the mass of IDS spiked into the sample and is a reflection of absolute mass recovery obtained during sample preparation and analysis steps. Except where specifically noted, all analyte recovery data are method recoveries obtained by using the isotope-dilution quantification procedure and, thus, are recoveries that are automatically corrected for procedural losses based on the absolute recovery of the IDS. The IDS's absolute recovery and the corresponding analyte's method recovery will not be the same, as detailed in this report.

Acknowledgments

The authors acknowledge the following current or former NWQL staff who provided assistance in the development and implementation of this method: Ed Furlong, Glenda Brown, Chris Kanagy, Donna Damrau, Burt Johnson, Serena Skaates, Jim Madsen, Jim Lewis, Bruce Darnel, Jeff McCoy, Barb Kemp, Tom Bushly, Mike Haschke, Craig Stappert, Duane Wydoski, Maggie Page, Joyce Wolff, Daniel Bizu, Phil Grano, Roy Brannan, Doug Mynard, Debbie Hobbs, and Susan McPartland. Thanks are extended to all USGS colleagues who submitted field samples and blanks during method testing, and to Patrick Phillips, Galen Hoogestraat, Kathy Lee, Melinda Erickson, and Steve Sando for also providing specific test matrices. The authors acknowledge the technical assistance on GC/MS/MS operation provided by Doug Stevens, Cari Randles, Blake McCurdy, Zack Blascak, and Steve Griffin of Waters Corporation. Edward Kolodziej of the University of Nevada–Reno, Douglas Latch of Seattle University, Jim Ounsworth of CDN Isotopes Inc., and Tom Dorsey of Cambridge Isotope Laboratories, Inc. are acknowledged for helpful discussions on deuterium isotope stability.

Colleen Rostad (USGS) and Glynda Smith (USEPA) provided technical review of this report; David Mueller (USGS) provided review of statistical applications.

Analytical Method

The analytical method for determination of selected steroid hormones and other compounds (table 1) in filtered water and unfiltered water is described in this section of the report. Steroid hormones and other compounds are analyzed using SPE and GC/MS/MS: (1) for filtered water using USGS method number O–2434–12, USGS method code GM004, NWQL laboratory schedule (LS) 2434, and (2) for unfiltered water using USGS method number O–4434–12, USGS method code GM005, NWQL LS 4434.

1. Scope and Application of Method

The method described in this report was developed by the USGS for use at the NWQL and was implemented at the NWQL on May 31, 2012 as an approved USGS method.

The method is designed for the determination of selected steroid estrogens, androgens, progestins, and related compounds in filtered (LS 2434) and unfiltered (LS 4434) water samples (table 1, fig. 1). Laboratory processing of samples for LS 2434 and LS 4434 at the NWQL is identical. The only difference between the schedules is whether or not samples are filtered (preferably in the field) before analysis. Many organic methods implemented at the NWQL provide determination of compounds from a wide variety of chemical classes. Conversely, most of the compounds determined in this method are structurally quite similar. Indeed, 18 of the method analytes share a common polycyclic steroid backbone and differ only in saturation or substitution. Bisphenol A and diethylstilbestrol, which themselves are structurally similar, do not share structural similarity with the steroid hormones but are known to act as endocrine system modulators (Mittendorf, 1995; Vandenberg and others, 2009).

The method is applicable to compounds that (1) are efficiently partitioned from water onto a C_{18} SPE disk, specifically an ENVI-18 disk (Sigma-Aldrich Corp., St. Louis, Mo.); (2) are effectively eluted from the SPE disk or collected particulate matter using methanol; (3) are not retained on Florisil when eluted with 5-percent methanol in dichloromethane; (4) possess functional groups with active hydrogens that can be derivatized using activated MSTFA; and (5) form trimethylsilyl ether or trimethylsilyl-enol ether derivatives that are stable and amenable to analysis using GC/MS/MS.

The method is applicable to both filtered and unfiltered water samples from a variety of matrices, including ground, surface, waste, and treated (including chlorinated) waters. Validation data have been obtained for the following water matrices: primary- and secondary-treated wastewater

effluents, wastewater-affected surface water, and reagent water. Although a statistically rigorous validation procedure has been undertaken only for the four matrix types listed above, more than 90 and 330 field samples were received for custom analysis by LS 2434 and 4434, respectively, during the method-validation period. These field samples represent a variety of matrices and sampling locations.

In addition to their role in the IDQ of analyte concentrations, the use of stable deuterium or ^{13}C-labeled compounds as IDS compounds in the method also provides insight to an analyte's absolute recovery in samples that have not been rigorously validated, because every sample analyzed is effectively a spike for the analytes that have exact IDS compounds (table 2; fig. 1). A summary of IDS recoveries in all custom samples analyzed in 2009–2010 is provided in this report. The IDS recoveries from these samples provide additional information about method performance in surface water, groundwater, agricultural runoff, combined-sewer overflow discharges, treated waters, and other matrices similar to those used in validation. Therefore, customers are not required to provide matrix-matched spike-recovery samples with all sample submissions; however, a customer QA plan that includes laboratory or field-matrix spikes, field blanks, and replicates is warranted.

Compounds determined by the method whose long-term recovery and variability fall within the criteria for acceptable performance are reported without qualification (NWQL Standard Operating Procedure (SOP) MX0015.x, "Guidelines for method validation and publication" (William Foreman and Robert Green, U.S. Geological Survey, written commun., 2005)). Concentrations for two compounds, equilin and progesterone, are reported as estimated ("E" remark code in NWIS; U.S. Geological Survey, 2011a) because their method performance is more variable than the other compounds analyzed by this method. Additional database coding is applied to other reported analyte data, as needed, based on sample-specific IDS recovery performance as described in section 12 of this report.

2. Summary of Method

Field samples are collected using USGS protocols for organic contaminants (section 5.6.1.F of Wilde and others, 2004), except that the samples are contained in 0.5-liter (L) high-density polyethylene bottles. Samples suspected to contain residual chlorine include 50 milligrams (mg) of ascorbic acid in the sample bottle. Samples not processed within 3 days at the NWQL are held frozen at –5 degrees Celsius (°C) or lower until the day preceding extraction, then allowed to thaw at room temperature. Field-filtered (LS 2434) and unfiltered (LS 4434) water samples are fortified with 50 mg of sodium chloride and the deuterium- and ^{13}C-labeled compounds that are used as isotope-dilution standards (table 2). The sample is extracted by solid-phase extraction by passing it through a multigrade glass-fiber filter (GFF) positioned over a glass-fiber filter disk that is embedded with octadecyl surface-modified silica (C_{18} disk). Following compound isolation, the GFF/C_{18} disk is rinsed with 10 milliliters (mL) of 25-percent methanol in reagent water to remove polar compounds that interfere with instrumental analysis. Nitrogen gas (N_2) is passed through the GFF/C_{18} disk to remove residual water, and the method compounds are eluted with 40 mL of methanol. The eluent is evaporated to dryness at 25°C with N_2 and reconstituted in 2 mL of a 5-percent methanol in dichloromethane solution.

The extract is transferred to a 1-gram (g) Florisil SPE column, and the analytes are eluted with 25 mL of 5-percent methanol in dichloromethane solution. The eluent is reduced in volume, transferred to a 5-mL reaction vial, and evaporated to dryness with N_2. Processing of multi-level calibration standards contained in reaction vials is included beginning at this evaporation step. Alcohol (C–OH) and ketone (C=O) groups on the analytes and IDS compounds are derivatized to trimethylsilyl ether or trimethylsilyl-enol ether analogs, respectively, to increase compound volatility and minimize compound interactions with active sites in the gas-chromatography system. Derivation is accomplished by (1) addition of 200 microliters (μL) of N-methyl-N-(trimethylsilyl)-trifluoroacetamide activated with 2-(trimethylsilyl)ethanethiol and ammonium iodide to the dried extract, and (2) heating of the MSTFA solution to 65°C for 1 hour (h). The MSTFA injection internal standard mixture also contains cholestane-2,2,3,3,4,4-d_6 (cholestane-d_6) and chrysene-d_{12} as injection internal standards.

This reconstituted extract is transferred to a vial for analysis. Analytes are separated by gas chromatography and detected by tandem-quadrupole mass spectrometry by monitoring the product ions of three specific precursor-to-product ion transitions as detailed in section 10. Positive analyte identification requires the presence of at least two unique transition product ions, with ion ratios not deviating from those in a standard by more than specified tolerances (Antignac and others, 2003); see section 10.8.3 for additional details regarding ion ratios.

All 20 method analytes are quantified relative to a specific IDS by using an isotope-dilution quantification procedure that automatically corrects for procedural losses in the reported analyte concentration by correction relative to the absolute recovery of the IDS. Thirteen deuterium- or ^{13}C-labeled isotopes currently (March 2012) are available that were found suitable for use as IDS compounds (fig. 2), 10 of which are exact isotopic analogs of the method analytes. Note: the exact isotopic analogs cis-androsterone-d_5, estriol-d_4, and progesterone-$^{13}C_3$ became commercially available since obtaining the validation data presented in this report, and they have been added to the method as the IDS compound used for the quantification of cis-androsterone, estriol, and progesterone, respectively. 16-Epiestriol-d_2 and medroxyprogesterone-d_3 that were used as IDS compounds for the estriol and progesterone data presented in this report were

subsequently changed to surrogate compounds as detailed in section 10.7 "Use of Isotope-Dilution Standards."

The remaining method analytes also are quantified using isotope dilution by using one of the IDS compounds that has similar related chemical functionality but is not a direct isotopic analog of the analyte. The IDS compounds are reported (in percent recoveries) along with the analyte concentrations to the NWIS database. However, these IDS measurements reflect absolute recoveries achieved during sample preparation and are only corrected for injection variability by quantitation by using the injection internal standards compounds chrysene-d_{12} or cholestane-d_6.

3. Health, Safety, and Waste-Disposal Information

All steps in the method that require the use of organic solvents, neat analytes, analytical standards, and silanization or derivatization reagents are conducted in a ventilated fume hood. Eye protection, gloves, and protective clothing are worn in the laboratory area and when handling standards, reagents, and solvents. Some of the reagents and compounds are, or are suspected to be, human carcinogens, teratogens, or mutagens. Many of the hormones are endocrine system modulators. It is important that protective measures are taken to avoid both dermal and respiratory exposure to solvents, reagents, standards, and sample extracts.

Protective gloves are worn at all times to minimize personal and inadvertent sample contamination. Glove protection varies based on glove type and chemical used. Silver Shield® gloves (North Safety Products, Cranston, R.I.) provide superior protection against dichloromethane (DCM) penetration, and their use is warranted when pouring or rinsing glassware with DCM, although manual dexterity is limited. Silver Shield®, Viton®, 4H® and butyl rubber and some types of rubber-based gloves are noted as protective against MSTFA dermal exposure (Merck Chemicals Company, 2010; Regis Technologies, Inc., 2010). Although nitrile gloves provide better dexterity, they are only splash resistant to DCM and MSTFA and, if used, are removed immediately if exposed to DCM, silanization, or derivatization reagents used in this method. Copies of Material Safety Data Sheets for the relevant reagents and analytes are reviewed before the use of the method. Some Material Safety Data Sheets are available at *http://www.ilpi.com/msds/index.html* (accessed April 2012).

All extracts in this method will contain one or more flammable solvents, and, when refrigerated, extracts are stored in flammable material or explosion safe refrigerators or freezers. Disposal of materials is carried out in strict accordance with current, local waste-handling regulations. Manufacturer's Material Safety Data Sheets are consulted for additional guidance on handling precautions, spill-cleanup procedures, and disposal of solvents, analytes, sample containers, or other materials used in this method. The NWQL's Safety, Health and Environmental Compliance

Section is the principal source for instructions regarding current waste handling procedures at the NWQL.

Samples that originate from locations affected by human or animal waste (wastewater treatment plants, septic systems, animal feeding operations) may contain possible biohazards. Wilde and others (2004) provide guidelines for safe collection and processing of these types of samples.

4. Interferences and Sample Contamination

Sample contamination is a concern because some of the hormone compounds are biogenic and can be present on human skin or might be used as personal-care products. Others analytes are synthetic hormone pharmaceuticals that are in common use. It is important that field and laboratory personnel exercise care to avoid contamination of the samples by avoiding consumption or contact with such materials immediately before and during sample collection and processing procedures. Excercising care is important for both the acquisition and subsequent handling of samples and sample extracts to avoid contamination. Collection of field blanks is essential to monitor for contamination; see section 13.4.1 for additional details regarding field blanks. Protective gloves (for example, nitrile gloves) must be worn at all times to minimize risk of sample or extract contamination.

Additionally, three compounds in the method are ubiquitous low-level procedural contaminants. Samples are easily contaminated with bisphenol A, 3β-coprostanol, and especially cholesterol during sample collection and preparation. Cholesterol is a primary sterol produced by animals that, contrary to widespread misinformation stating otherwise, also is produced by and present in plants, although at substantially lower concentrations than in animals (Behrman and Gopalan, 2005). It is important to avoid laboratory tissue or other paper products during field or laboratory sample manipulations of these samples because these products might contain cholesterol. More importantly, cholesterol is present in human skin flakes and occurs at substantial concentrations in both indoor and outdoor dust (Weschler and others, 2011). The analyte 3β-coprostanol is a fecal sterol that, coupled with other sterol concentration information, is used as an indicator of human fecal-waste sources (Leeming and others, 1996). Likewise, 3β-coprostanol is associated with dust, which might be a potential contamination source when collecting water samples downwind from biosolids-treated fields or animal feeding operations. Bisphenol A is used primarily in the manufacture of polycarbonate plastic and epoxy resins, and both of these product types have many commercial applications that might be encountered in either the laboratory or field environment (see, for example, PlasticsEurope, 2007). Data for these three blank-limited compounds are reported using the minimum reporting level (MRL) convention (Childress and others, 1999).

A siloxane compound thought to emanate from the GC column or vial septum at low levels (less than (<) 1 ng/L)

interferes with the precursor ion used for determination of 11-ketotestosterone (also reported using the MRL convention; see the "Blank-Limited Analytes" section). Sample-specific matrix interferences might warrant the application of a raised reporting level value for an analyte or the use of other accepted data qualifiers, such as the NWIS estimated (E) remark code or deletion codes. See the "Assessment of Blank Contamination and Determination of Method Detection and Reporting Levels" section for more information.

5. Apparatus and Instrumentation

The apparatus and instrumentation used for the method are outlined in this section, and, except as noted, are grouped using the same subsection heading name that is used in section 9 "Sample Preparation," where the apparatus or instrumentation is first used during sample preparation. Alternative apparatus and instrumentation from those listed for this method may be substituted if shown, or known from the literature, to provide comparable or superior performance and analyte recoveries. Therefore, the phrase "or equivalent" is not included for the item descriptions in this report. Some materials are common laboratory items, and, therefore, are not described in detail.

5.1. Cleaning of General Glassware

5.1.1. *Dishwasher.*
5.1.2. *Oven:* programmable and capable of heating to at least 450°C for 2 h.

5.2. Cleaning and Silanization of Specific Glassware

5.2.1. *50-mL Glass receiver tube*: custom-fabricated tube (Allen Scientific Glassware, Inc., Boulder, Colo., catalog number (cat. no.) ASG–50–SPT) by adding a Pyrex® 24/40 ground-glass opening to the top of a 50-mL conical, graduated centrifuge tube (Corning, Inc., Corning, N.Y., cat. no. 8080–50). This glassware is to be used only for hormone methods.
5.2.2. *40-mL SPE receiver tube*: custom-fabricated Pyrex® glass tube (Allen Scientific Glassware, Inc. cat. no. ASG–RT–H), 13.3 centimeters (cm) long and a 2.8-cm outer diameter (o.d.) with 19/22 ground-glass opening and bottom nipple taper with 0.5-mL volume calibration mark. This tube is decal-labeled "H" for use as hormone method-specific glassware to facilitate segregation from identical non-silanized tubes used in other methods.
5.2.3. *Glass reaction vial*: 5-mL, V-bottom, 20-millimeter (mm) o.d., with 20–400 thread open-top cap and polytetrafluoroethylene-faced septum (Sigma-Aldrich Corp., St. Louis, Mo., cat. no. 33299). This glassware is to be used only for hormone methods.

5.2.4. *Oven*: programmable and capable of heating to at least 450°C (Lindberg/Blue M model BF51828C–1, Thermo Fisher Scientific, Inc., Waltham, Mass.).

5.3. Sample Thawing and Set Creation

5.3.1. *Refrigerator*, if needed.

5.4. Extractor Cleaning

The extractor consists of the following:

5.4.1. *Stainless-steel extractor (SSE)*: a custom-fabricated 530-mL capacity, pressurized extractor (Martin Enterprises, Lakewood, Colo., cat. no. SSFT500mlSet) consisting of the following components (fig. 3):

5.4.1.1. *Stainless-steel (type 304) tube body*: 35.56-cm long by 5.08-cm o.d. (4.5-cm inner diameter (i.d.)) (Martin Enterprises, cat. no. SSFT500ml).

5.4.1.2. *Bottom end cap* (fig. 3):, fabricated from type 303 stainless steel and machined on the interior (Martin Enterprises, cat. no. BEC47mm; fig. 4*A*) to contain a 3.5-cm diameter by 0.2-cm thick stainless-steel frit (60-micrometer (μm) nominal pore size; Martin Enterprises, cat. no. SSSD1.375x60), a C_{18} disk (see section 6.5), and glass-fiber filter (see section 6.5), all held in place with a 4.75-cm o.d. (4.25-cm i.d.) by 0.13-cm thick Teflon® back-up O-ring (Martin Enterprises, cat. no. TBR8–223). The bottom end cap also is fitted with an 8.89-cm long by 0.95-cm o.d. (0.76-cm i.d.) drain tube (Martin Enterprises, cat. no. DT0.375) using Teflon® tape on the tube's 0.3175-cm (1/8-inch) female pipe thread screw threads. The bottom end cap is screw-thread connected to the bottom thread of the stainless-steel tube body with a 5.08-cm i.d. by 0.19-cm thick Viton® O-ring (Martin Enterprises, cat. no. OR2–032 Viton), and, when tight, the

bottom end of the SSE tube seals against the Teflon® back-up O-ring in the cap (see figs. 3 and 4*B*).

5.4.1.3. *Top end cap*: fabricated from type 303 stainless steel and thread connected to the stainless-steel tube body (Martin Enterprises, cat. no. TEC47mm) with a 5.08-cm i.d. by 0.19-cm thick silicone O-ring (Martin Enterprises, cat. no. OR2–032 Silicone), and fitted with a 0.3175-cm National Pipe Taper Thread, quick-connect body (Swagelok Company, Solon, Ohio, cat. no. B–QM2–S–200).

5.5. Extractor Assembly (Pre-cleaned)

The extractor is assembled using the following:

5.5.1. *Manifold panel rack*: a custom-fabricated rack (Martin Enterprises, cat. no. MPRack6; fig. 3) designed to hold six stainless-steel extractors and to supply a controlled flow of nitrogen gas to each SSE. Connecting two racks to the gas source provides capacity for 12 SSEs within a large hood. Main rack components include the following:

5.5.1.1. *Six SSE clamp holders*, each tightened against the SSE using a screw and wing nut.

5.5.1.2. *Gas delivery manifold* consisting of variable lengths of 0.3175-cm o.d. by 0.16-cm i.d. tetrafluoroethylene Teflon® tubing (Sigma-Aldrich Corp., cat. no. 58699), and connected with the 0.3175-cm (1/8-inch) i.d. components: three union crosses (Swagelok Company, cat. no. B-200-4), six 3-way valves (Swagelok Company, cat. no. B-41XS2), and six needle valves (Swagelok Company, cat. no. B–ORS2). Flow of N_2 (provided through a NWQL-wide gas delivery system) to the entire manifold is controlled by a 2-stage gas regulator. Connection of Teflon® tubing from outlet of the needle valve to the quick-connect body in the top end cap of the SSE is accomplished with a quick-connect stem (Swagelok Company, cat. no. B–QM2–S–200). During operation, application and relief of gas pressure to the SSE is accomplished by switching of the 3-way valve. Additional flow control is accomplished with both the 2-stage regulator and the SSE-specific needle valve (fig. 3; note: some components listed in this section are not shown in the figure).

5.6. Quality-Control Sample Preparation and Isotope-Dilution Standard Addition

5.6.1. *Water-purification system*: Solution 2000 system (Aqua Solutions, Incorporated, Jasper, Ga., model 2002AL).

5.6.2. *Balance*: top-loading, capable of weighing to at least 1,000 g with an accuracy of plus or minus (±) 0.1 g (Mettler-Toledo, Columbus, Ohio, model PB3002).

5.6.3. *IDS 100-μL microdispenser (for IDS addition only)*: fixed volume, glass capillary-type with flouropolymer plunger tip (VWR International, Radnor, Pa., cat. no. 53506–653). Note: Dispenser is labeled for isotope-dilution standard addition only and stored separately from dispensers or syringes used for dispensing method analytes.

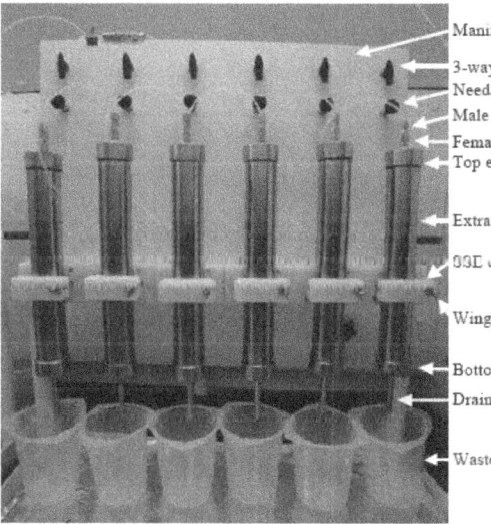

Manifold panel rack

3-way valve
Needle valve
Male quick connector
Female quick connector
Top end cap

Extractor tube body

SSE clamp holder

Wing nut

Bottom end cap
Drain tube

Waste beaker

Figure 3. Stainless-steel extractor (SSE) tubes mounted on a 6-position manifold panel rack.

 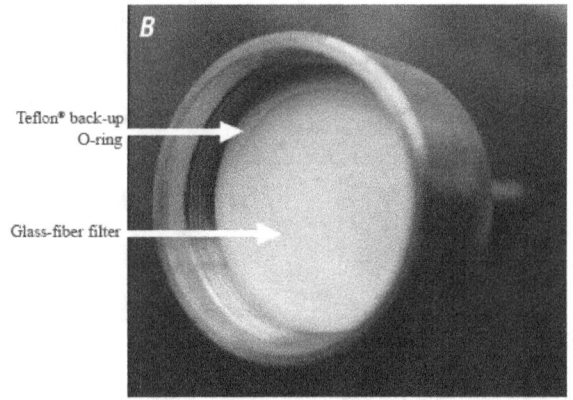

Figure 4. Bottom end cap (*A*) with machined interior and (*B*) after installation of stainless-steel frit (not shown), C$_{18}$ disk (not shown), glass-fiber filter (shown), and Teflon® back-up O-ring (shown).

5.6.4. Spike 100-µL microdispenser (for spike mixture addition only): fixed volume, glass capillary-type with flouropolymer plunger tip (VWR International, cat. no. 53506-675). Note: Dispenser is labeled for hormone spike mixture addition only (section 7.1.7) and stored separately from dispensers or syringes used for dispensing IDS or MSTFA/internal injection standard (MSTFA/IIS) solutions.

5.6.5. Scoopula: stainless steel (Thermo Fisher Scientific, Inc., cat. no. 14-357Q).

5.7. Sample Extraction

5.7.1. Plastic waste beakers: 1-L beakers to collect solvent rinses and extracted water from SSE.

5.8. Elution of GFF/C$_{18}$ disk

5.8.1. 50-mL Glass receiver tube: specified in section 5.2.1. The tube is cleaned and silanized as described in section 9.2 before use in section 9.8. This glassware is to be used only for hormone methods.

5.8.2. 250-mL Glass beaker: beaker for holding the 50-mL glass receiver tube.

5.8.3. 24/40 Ground-glass stopper: stopper (Thermo Fisher Scientific, Inc., cat. no. 14–640–J).

5.8.4. Rack: 24-position polypropylene rack for 25–30-mm o.d. tube (VWR International, cat. no. 60985–228).

5.9. First Evaporation

5.9.1. Nitrogen evaporator: 24-position evaporator with thermostatically-controlled water bath (N-Evap, Organomation Associates Inc., Berlin, Mass.), supplied with the NWQL's nitrogen gas delivery system purified by passing through an activated charcoal trap for organic contaminant removal.

5.9.2. Needles: 10.2-cm long, 19-gage stainless-steel needles with Luer-lock fitting. Needles are cleaned between uses by sonication with methanol.

5.10. Florisil Cleanup of Extract

5.10.1. 40-mL SPE receiver tube: custom-fabricated tube (Allen Scientific Glassware, Inc., cat. no. ASG–RT–H) 13.3-cm long by 2.8-cm o.d. Pyrex glass tube with 19/22 ground-glass opening and bottom nipple taper with 0.5-mL volume calibration mark. This tube is decal-labeled "H" as hormone method-specific glassware to facilitate segregation from identical non-silanized tubes used in other methods. Only receiver tubes that have been cleaned and silanized as described in section 9.2 are used. This silanized tube is used only for hormone methods.

5.10.2. 24/40 Ground-glass stopper.

5.10.3. Visiprep™ DL 24 vacuum manifold: for solid-phase extraction (Sigma-Aldrich Corp., cat. no. 57265), includes a top identified for use with hormone SPE cleanup only and collection rack modified to hold six 40-mL SPE receiver tubes (see fig. 5 in section 9).

5.10.4. Air-driven vacuum pump, with connecting tubing (Sigma-Aldrich Corp., cat. no. 506192). The pump's air inlet is connected with tubing to the NWQL's nitrogen gas delivery system. Tubing from the vacuum inlet on the pump is routed through a 20-L solvent-waste carboy, by using fittings on the carboy's cap, to the Visiprep DL 24 vacuum manifold.

5.10.5. Vortex mixer: mixer (Scientific Industries, Inc., Bohemia, N.Y., model G–560).

5.11. Second Evaporation

5.11.1. See section 5.9.

5.12. Transfer Extract to Reaction Vial

5.12.1. *Vial rack*: 36 position with 24-mm opening for use with reaction vials (Sigma-Aldrich Corp. cat. no. 23201).

5.13. Preparation of Calibration Standards

5.13.1. *Syringes*: 25 and 100 µL. The syringes are labeled specifically for use with Hormone Analyte Mix solutions only for calibration standard preparation. In particular, these syringes are not used for the preparation of intermediate stock solutions for the hormone method to avoid contaminating these solutions.

5.14. Third Evaporation

5.14.1. See section 5.9.

5.15. Derivatization and Extract Transfer to GC Vial

5.15.1. *200-µL Microdispenser (for MSTFA/IIS addition)*: fixed volume, glass-capillary type with flouropolymer plunger tip (VWR International, cat. no. 53506-697). Note: Dispenser is labeled for MSTFA/IIS addition only and stored separately from dispensers or syringes used for dispensing IDS compounds or for spiking method analytes.
5.15.2. *Heating block*: 12-position (Barnstead Thermolyne, Type 17600 Dry Bath, Thermo Fisher Scientific, Inc.) or 24-position heating block (Barnstead Labline Dry Bath, cat. no. 2053Q, Thermo Fisher Scientific, Inc.).
5.15.3. *Thermometer*: alcohol-based bulb type, 0–100°C range with 1°C gradation.
5.15.4. *GC vial rack*: rack (Sigma-Aldrich Corp., cat. no. 23207).

5.16. Standards

The items below are used in section 7 "Standards."
5.16.1. *Class A volumetric glassware and pipettes*: various volumes as needed.
5.16.2. *Syringes*: variable volumes, with fluoropolymer plunger tip (for example, maximum calibrated ranges from 10 to 500 µL). Note: syringes with cemented needles have been used for this method. However, Watabe and others (2004) have noted that syringes with removable needles might provide lower potential for bisphenol A contamination; this type of syringe has not been tested.
5.16.3. *Balance*: capable of weighing to ±0.01 mg (Mettler-Toledo model XS205).
5.16.4. *Vials*: variable volumes ranging from 5–60 mL with polytetrafluoroethylene-lined lids for storage of standards.

5.17. Analytical Instrumentation and Computer Hardware/Software

The items below are used in section 10 "Analysis by Gas Chromatography with Tandem Mass Spectrometry."
5.17.1. *GC*: Agilent 6890 gas chromatograph with 7673B autosampler (Agilent Technologies, Santa Clara, Calif.).
5.17.2. *GC/MS/MS*: quadrupole-hexapole-quadrupole tandem mass spectrometer (Waters Corp., Milford, Mass., Quattro-micro-GC™).
5.17.3. *Personal computer*: running Microsoft® Windows XP.
5.17.4. *MassLynx™ operational and TargetLynx™ data processing software*: version 4.1 or later (Waters Corp.).
5.17.5. *MaDCU*: a data processing program written in Visual Basic by Chris Lindley at the NWQL that performs additional calculations and data formatting for input into the NWQL's laboratory information management system and, ultimately, into NWIS (NWQL SOP ORGM0477.x, "Analysis of hormone samples by GC/MS/MS—Laboratory schedules 2434, 4434, 6434, and 7434," Chris Lindley and others, written commun., 2011).

6. Reagents and Consumables

The reagents and consumables used in the method are outlined in this section, and, except as noted, are grouped using the same subsection heading name that is used in section 9 "Sample Preparation" where the reagent or consumable is first used during sample preparation. Alternative reagents and consumables may be substituted if shown, or known from the literature, to provide comparable or superior performance and analyte recoveries. Therefore, the phrase "or equivalent" is not included for the item descriptions in this report. Some materials are common laboratory items and, therefore, not detailed. Unless otherwise specified, solvents used for rinsing typically are contained in fluorinated ethylene propylene (Teflon® FEP) wash bottles with ethylene-tetrafluoroethylene (Tefzel® ETFE) stem closure and tubing. All solvents are pesticide-residue grade or better.

6.1. Cleaning of General Glassware

6.1.1. *Neutrawash® detergent* (Getinge USA, Inc., Rochester, N.Y. cat. no. 61301600011).
6.1.2. *Aluminum foil*.

6.2. Cleaning and Silanization of Specific Glassware

6.2.1. *Methanol*.
6.2.2. *Liquinox® detergent solution*: prepared by mixing about 2 drops of Liquinox® (Alconox, Inc., White Plains, N.Y.) with about 500 mL of hot tap water.
6.2.3. *Brushes*: various types for bottles and test tubes.

6.2.4. *5-Percent dimethyldichlorosilane in toluene*: Sylon CT™ (Sigma-Aldrich Corp., cat. no. 33065–U).

6.3. Sample Thawing and Set Creation

6.3.1. None.

6.4. Extractor Cleaning

6.4.1. Tap water, deionized water, methanol, acetone, and hexane.

6.5. Extractor Assembly

6.5.1. *Nitrogen gas*: delivered through NWQL-wide distribution system from a liquid nitrogen tank.

6.5.2. *Multigrade glass-fiber filter (GFF)*: 47-mm diameter, graded from 25- to 1-μm nominal porosity across the filter face based on water flow path (Whatman Incorporated, Piscataway, N.J., GMF150, cat. no. 1841–047; Thermo Fisher Scientific, Inc., cat. no. 09–874–82).

6.5.3. *Supelclean™ ENVI™-18 SPE Disk (C$_{18}$ disk)*: 47-mm diameter extraction disk containing reverse phase octadecyl surface-modified-silica embedded glass-fiber filter disk with 5-μm mean flow-through porosity (Sigma-Aldrich Corp., cat. no. 57171).

6.5.4. *Teflon® (polytetrafluoroethylene) tape* 1.27-cm width.

6.6. Quality-Control Sample Preparation and Isotope-Dilution Standard Addition

6.6.1. *Reagent water*: typically prepared by using Solution 2000 water-purification system (see section 5.6.1) or obtained as bottled, pesticide-grade water supplied for use for organic analyses by the USGS National Field Supplies Service cat. no. N1580 or N1590; contact information available at *http://nwql. usgs.gov/about-contacts.shtml*.

6.6.2. *Sample bottle*: 0.5-L high-density polyethylene (HDPE) bottles (USGS National Field Supplies Service cat. no. Q36FLD).

6.6.3. *Polyethylene-lined, polypropylene screw cap*: 28-mm thread (USGS National Field Supplies Service cat. no. Q417FLD).

6.6.4. *Glass bores*: 50- and 100-μL glass bores for micro dispensers. Baked at 450°C for 2 h before use.

6.6.5. *Salt*: sodium chloride (NaCl); American Chemical Society reagent-grade salt (J.T. Baker, cat. no. 3624–07; Anvantor Performance Material, Phillipsburg, N.J.); precleaned by baking for 4 h at 450°C and stored in a capped jar.

6.7. Sample Extraction

6.7.1. *Methanol*: contained in an 1-L amber, glass bottle with 10-mL bottle-top dispenser (Brinkmann product no. 022220209, Metrohm USA Incorporated, Riverview, Fla.).

6.7.2. *Reagent water*: prepared by using Solution 2000 water-purification system (see section 6.6.1), and contained in a 4-L amber glass bottle with 25-mL bottle-top dispenser (VWR International, cat. no. 40000–066).

6.7.3. *25-Percent methanol/water solution*: prepared by volume using 250 mL of methanol and 750 mL of reagent water (see section 6.6.1). The solution is contained in an 1-L amber, glass bottle with 10-mL bottle-top dispenser (Brinkmann, product no. 022220209). Note: solution is mixed slowly to minimize exothermic reaction.

6.8. Elution of GFF/C$_{18}$ disk

6.8.1. *Methanol*: see section 6.7.1.

6.9. First Evaporation

6.9.1. *Nitrogen gas*: delivered through NWQL-wide distribution system; includes in-line hydrocarbon trap.

6.10. Florisil Cleanup of Extract

6.10.1. *5-Percent methanol in dichloromethane solution (5-percent methanol/DCM)*: prepared by volume (for example, if 1-L amount is needed, 50-mL methanol and 950-mL DCM are used) and contained in a 1-L amber, glass bottle with a 5-mL bottle-top dispenser (Brinkmann, product no. 022220101).

Note: Mixture may be stored sealed for as long as 1 month. Between sample set preparation, the bottle-top dispenser is removed and screwed onto an empty, clean bottle for dispenser storage. The bottle containing the 5-percent methanol/DCM solution is capped; this minimizes selective volatilization of DCM relative to the methanol (which alters the intended composition) if the dispenser is left on the bottle containing the mixture for extended periods of time.

6.10.2. *Florisil SPE column*: 1-g sorbent, 6-mL barrel (Biotage, LLC, Charlotte, N.C., cat. no. 712–0100–C).

6.10.3. *Teflon® valve liners*: for the Visiprep™ DL 24 vaccuum manifold (Sigma-Aldrich Corp., cat. no. 57059).

6.10.4. *Pasteur pipettes*: 6- or 9-cm long with latex rubber bulbs.

6.11. Second Evaporation

6.11.1. See section 6.9.

6.12. Transfer Extract to Reaction Vial

6.12.1. *Glass reaction vial*: 5-mL, V-bottom, 20-mm o.d., with Glass Packing Institute thread 20–400 open-top cap and polytetrafluoroethylene-faced septum (Sigma-Aldrich Corp., cat. no. 33299). The vial is cleaned and silanized as described in section 9.2.
6.12.2. *Septum*: polytetrafluoroethylene-faced silicone, 20-mm diameter, for use with reaction vials (Wilmad-Lab Glass, Vineland, N.J., cat. no. LG–4342–108).
6.12.3. *Cap*: black phenolic, open top, 20–400 thread (Wilmad-Lab Glass, cat. no. LG–4341–108).

6.13. Preparation of Calibration Standards

6.13.1. *Low (0.01 nanogram per microliter [ng/μL]) and high (1 ng/μL) hormone analytes mixtures*: See sections 7.1.5–7.1.6 for mixing details.

6.14. Third Evaporation

6.14.1. See section 6.9.

6.15. Derivatization and Extract Transfer to GC Vial

6.15.1. *MSTFA (activation II) injection internal standard (IIS) mixture (MSTFA/IIS)*: see section 7.1.8.
6.15.2. *Glass bores*: 200-μL for micro dispenser.
6.15.3. *GC vial*: 12-mm o.d. by 32-mm long with 300-μL sealed insert, clear, 9-mm ABC™ screw-thread (Wheaton Science Products, Millville, N.J., cat. no. 225326; VWR, cat. no. 16150–410).
6.15.4. *GC vial cap*: 9-mm screw-thread cap (Wheaton Science Products, cat. no. 225333–03SP; VWR, cat. no. 14213–324).

6.16. Standards

The following are used in section 7 "Standards."
6.16.1. *Vials*: 2 to 60-mL range, glass, Teflon®-faced silicone rubber-lined screw caps.

6.17. Analysis by Gas Chromatography with Tandem Mass Spectrometry

The following are used in section 10 "Analysis by Gas Chromatography with Tandem Mass Spectrometry."
6.17.1. *Helium*: 99.999 percent purity (General Air Service and Supply, Denver, Colo., cat. no. 1214703000).
6.17.2. *Argon*: gas supply provide by NWQL distribution system from bulk, cryogenic argon tank (General Air Service and Supply, item. no. 17).

6.17.3. *GC injection-inlet liner*: Siltek®-deactivated, single gooseneck liner with 4-mm i.d. (Restek Corporation, Bellefonte, Pa., cat. no. 20800-214.25).
6.17.4. *Rxi®-XLB capillary GC column*: 30-meter long by 0.25-mm internal diameter with 0.25-μm film thickness (Restek Corporation, cat. no. 13726).

7. Standards

The preparations of standards used in the method are outlined in this section.
Glassware Note: Class A volumetric flasks and pipettes are cleaned by thorough rinsing with solvents and allowed to air dry briefly; this glassware is not baked in an oven. The 5-mL reaction vials used to prepare instrument calibration standards must be cleaned and the interior surfaces deactivated by silanizing using the procedures described in section 9.2. Other glassware is cleaned using a dishwasher and baked using an oven temperature cycle that includes maintaining an upper temperature of 450°C for at least 2 h.
Restricted analytes Note: In the United States, storage and handling of 4-andostene-3,17-dione, 11-ketotestosterone, dihydrotestosterone, testosterone, and nandrolone-d_4 are regulated by and require licensing from the U.S. Drug Enforcement Administration (see *http://www.deadiversion.usdoj.gov/*).

7.1 Preparation of Standard Solutions

The solution compositions, concentrations, and fortification quantities (for analytes and IDS compounds and the MSTFA/IIS solution) given in the following subsections were those used during method validation or custom implementation. Alternative solution compositions can be used based on need, or solution or analyte availability. The fortification solutions are prepared as dilutions of higher concentration intermediate standards that are prepared by NWQL staff, or commercial vendors as appropriate, using neat standards obtained commercially. Additional details regarding preparation, verification, and documentation requirements are provided in NWQL SOP ANLX0478.x, "Documentation, verification, and labeling of standard solutions/materials, reference materials, solvents, and reagents" (Duane Wydoski and others, written commun., 2011).
7.1.1. *Storage and use*: Fortification solutions are prepared as detailed in this section and verified in advance of their use. After formulation, all standard solutions are stored frozen (<–5°C), unless otherwise noted, in glass vials with Teflon®-faced, silicone rubber-lined screw caps. Before addition to samples, stock and fortification solutions are removed from the freezer and allowed to reach room temperature. The solutions are thoroughly mixed (by shaking or using a vortex mixer) before use to ensure homogeneity. The solutions are returned to freezer storage immediately after use. Sub-aliquots of prepared fortification mixtures are stored in 2- or 4-mL

vials to minimize solvent volatilization and repeated warming of an entire prepared volume of the solution during subsequent usages. Unless noted otherwise, standard solutions are used for no more than 1 year before recertification is required to validate concentrations or new verified solutions are prepared. Old solutions typically are retained for an additional year beyond the solution expiration date to allow for future comparisons with newer standards, if needed, should problems arise with determined concentrations or stability of newer standards.

7.1.2 *Stock standards and solutions*: Individual stock standard solutions are prepared from high purity (greater than or equal to ($\geq$) 97.5 percent for analytes or $\geq$95 atom percent deuterium or ^{13}C for IDS compounds) solid standards. Commercial sources of neat solid standards used during method validation are given in table 3. These solid standards are stored as recommended by the manufacturer.

The target concentration for individual stock standard solutions is 3 micrograms per microliter ($\mu g/\mu L$), but may vary from 1–10 $\mu g/\mu L$ depending on the exact mass of solid material weighed out. Approximately 15 mg of the solid analyte material is accurately weighed to the nearest 0.01 mg and dissolved in 5 mL of methanol (milligrams per milliliter is equivalent to $\mu g/\mu L$). (Note: Stock solutions of the IDS compounds are prepared in acetone instead of methanol; toluene is used instead of acetone as the diluent for cholestane-d_6). Some compounds (particularly cholesterol and 3β-coprostanol) will be near their maximum solubility in methanol in these concentrated standards and an alternative solvent(s) can be used. When standard solutions are prepared, and before each use, it is important to ensure that all solid material is completely dissolved. Sonication or gentle heating (not above 50°C) may be applied to complete dissolution of analytes.

7.1.3. *Intermediate mixed standards* (100 ng/μL or other concentration, as needed): Intermediate mixtures of the 20 method analytes are prepared in methanol by dilution mixing of stock standards. The following three distinct intermediate standards are prepared: (1) contains cholesterol and 3β-coprostanol, (2) contains bisphenol A only, and (3) contains the other 17 analytes. For intermediate IDS standard solutions, one with cholesterol-d_7 only and one with the other IDS compounds are prepared. Note: intermediate IDS solutions are prepared in acetone. Cholestane-d_6 is prepared in toluene.

7.1.4. *Isotope-dilution standard fortification mixture (IDS mixture)*: Prepared by dilution of the intermediate IDS solution into acetone to yield the concentrations shown in table 3. For simplicity during data analysis, all of the IDS compounds except cholesterol-d_7 typically are included in this solution at the same concentration (0.5 ng/μL). Cholesterol-d_7 is included at 50 ng/μL because concentrations of cholesterol and 3β-coprostanol typically are higher than other analytes in samples. Beginning March 17, 2011, estriol-d_4 replaced 16-epiestriol-d_2 (which was changed to a surrogate compound) as the IDS for quantifying estriol. Beginning October 1, 2011, *cis*-androsterone-d_5 replaced nandrolone-d_3 as the

IDS for quantifying *cis*-androsterone. Beginning March 1, 2012, progesterone-$^{13}C_3$ replaced medroxyprogesterone-d_3 (which was changed to a surrogate compound) as the IDS for quantifying progesterone.

Before sample extraction, 100 μL of the IDS mixture is added to each field and QC sample, and to each calibration standard.

Note 1: Isotope-dilution quantification relies on use of the same solution lot number and same mass amount fortified of IDS compounds for all samples and calibration standards that are analyzed within a batch instrumental run (sequence). Use of different IDS lots having different IDS concentrations or different IDS fortification amounts in samples relative to the calibration standards in the batch run might lead to substantial bias in the determined analyte concentration in the sample.

Note 2: Six deuterated IDS compounds (4-androstene-3,17-dione-2,2,4,6,6,16,16-d_7, dihydrotestosterone-1,2,4,5a-d_4, estrone-2,4,16,16-d_4, norethindrone-2,2,4,6,6,10-d_6, testosterone-2,2,4,6,6-d_5, and progesterone-2,2,4,6,6,17a,21,21,21-d_9) initially tested and used in the method were found to undergo deuterium loss in methanol solution due to deuterium exchange with hydrogen even when the solution was mostly stored at less than −5°C (Foreman and others, 2010) (see section 10.7). These labile IDS compounds were eliminated from the method. The remaining deuterated IDS compounds did not exhibit deuterium loss. Nonetheless as an additional precaution, acetone is prescribed instead of protic solvents like methanol for preparation of the IDS stock, intermediate, and analyte standard solution mixtures.

7.1.5. *High-analytes mixture*: This mixture contains the 20 method analytes in methanol at the final concentrations used and shown in table 3. At least 25 mL of this mixture is prepared based on subsequent use requirements. The final concentrations of this mixture are 100 times greater than those in the low-analytes mixture.

7.1.6. *Low-analytes mixture*: Prepare by dilution of the high-analytes mixture into methanol to yield the final concentrations used and shown in table 3. At least 10 mL of this mix is prepared based on subsequent use requirements.

7.1.7. *Laboratory schedule 2434/4434 spike mixture* (2434/4434 spike mixture; nominal 0.125 ng/μL or other concentration, as appropriate): Prepared by dilution of the high-analytes mixture into methanol (typically 50 mL; split into 10-mL aliquots for storage). Before sample extraction, 100 μL of spike mixture is added to laboratory reagent water spike sample (also called "set" spike) and to any laboratory or field matrix-spike samples (see section 9.6).

Note on spike volumes: The volume of the 2434/4434 spike mixture added to laboratory matrix-spike samples (MSPK) or to field-requested laboratory matrix-spike samples (FRLMS) might be greater than the typical 100-μL based on anticipated ambient concentrations of analytes in the sample matrix.

NWIS note: The spike mixture used to fortify the FRLMS has an NWIS lot number assigned that provides access to

Table 3.   Compound name, Chemical Abstract Service (CAS) Registry Number, commercial source, catalog number, and concentration of compounds in the low- and high-analytes mixture standards or in the isotope-dilution standard fortification mixture (IDS mixture) used for calibration standard preparation.

[CDN, C/D/N Isotopes Inc. (*https //www.cdnisotopes.com/*); CIL, Cambridge Isotope Laboratories, Inc. (*http //www.isotope.com/cil/*); mL, milliliters; NA, not available; ng/µL, nanograms per microliter; NSI, NSI Solutions, Inc. (*http //www.nsi-es.com/*); PBI, Pfaltz and Bauer, Inc. (*https //www.pfaltzandbauer.com/*); Sigma, Sigma-Aldrich Company (*http //www.sigmaaldrich.com/*); TRC, Toronto Research Chemicals, Inc. (*http //www.trc-canada.com/*); --, not applicable]

Compound	CAS Registry Number	Commercial source	Catalog number	Low-analytes mixture (ng/µL)	High-analytes mixture (ng/µL)	IDS mixture (ng/µL)
Solution volume used				10 mL	25 mL	25 mL
Analytes						
11-Ketotestosterone	564–35–2	Sigma	K8250–5MG	0.01	1	--
17-*alpha*-Estradiol	57–91–0	Sigma	E8750–100MG	0.01	1	--
17-*alpha*-Ethynylestradiol	57–63–6	Sigma	E4876–1G	0.01	1	--
17-*beta*-Estradiol	50–28–2	Sigma	E8875–250MG	0.01	1	--
3-*beta*-Coprostanol	360–68–9	Sigma	C7578–50MG	1	100	--
4-Androstene-3,17-dione	63–05–8	Sigma	A9630–1G	0.01	1	--
Bisphenol A	80–05–7	Sigma	239658–50G	0.1	10	--
Cholesterol	57–88–5	Sigma	362794–5G	1	100	--
cis-Androsterone	53–41–8	Sigma	219010–1G	0.01	1	--
Dihydrotestosterone	521–18–6	Sigma	A8380–1G	0.01	1	--
Epitestosterone	481–30–1	Sigma	E5878–100MG	0.01	1	--
Equilenin	517–09–9	PBI	E01560–25MG	0.01	1	--
Equilin	474–86–2	Sigma	E8126–100MG	0.01	1	--
Estriol	50–27–1	Sigma	E1253–100MG	0.01	1	--
Estrone	53–16–7	Sigma	E9750–1G	0.01	1	--
Mestranol	72–33–3	Sigma	855871–500MG	0.01	1	--
Norethindrone	68–22–4	Sigma	N4128–1G	0.01	1	--
Progesterone	57–83–0	Sigma	P0130–25G	0.01	1	--
Testosterone	58–22–0	Sigma	T1500–1G	0.01	1	--
trans-Diethylstilbestrol[a]	56–53–1	Sigma	D4628–1G	0.01	1	--
Isotope-dilution standards (IDSs)						
17-*alpha*-Ethynylestradiol-2,4,16,16-d_4	350820–06–3	CDN	D–4319	--	--	0.5
17-*beta*-Estradiol-13,14,15,16,17,18-$^{13}C_6$	NA	CIL	CLM–7936–0.1MG	--	--	0.5
Bisphenol A-d_{16}	96210–87–6	CIL	DLM–1839–0	--	--	0.5
Cholesterol-25,26,26,26,27,27,27-d_7	83199–47–7	CDN	D–3557	--	--	50
cis-Androsterone-2,2,3,4,4-d_5[b]	NA	CDN	D–7185	--	--	0.5
Diethyl-1,1,1',1'-d_4-stilbestrol-3,3',5,5'-d_4[a]	91318–10–4	CDN	D–2849	--	--	0.5
Estriol-2,4,16,17-d_4[c]	NA	CIL	DLM–8583–0.1MG	--	--	0.5
Estrone-13,14,15,16,17,18-$^{13}C_6$	NA	CIL	CLM–7935–0.1MG	--	--	0.5
Mestranol-2,4,16,16-d_4	NA	CDN	D–6142	--	--	0.5
Nandrolone-16,16,17-d_3	120813–22–1	CDN	D–5735	--	--	0.5
Progesterone-2,3,4-$^{13}C_3$[d]	327048-87-3	Sigma	737143	--	--	0.5
Surrogates						
16-Epiestriol-2,4-d_2[e]	366495–94–5	CDN	D–5551	--	--	0.5
Medroxyprogesterone-20,20,20-d_3[f]	162462–69–3	TRC	M203552	--	--	0.5
Injection internal standards (IISs)						
Cholestane-2,2,3,3,4,4-d_6 {Cholestane-d_6}	358731–18–7	CDN	D–5535	--	--	--
Chrysene-d_{12}	1719–03–5	NSI	C–394L	--	--	--

[a]*trans*-Diethylstilbestrol and *trans*-diethylstilbestrol-d_8 are the only isomeric forms of these compounds that are being determined by the method. The *cis*-isomer forms are present in the standard material at less than 7 percent.

[b]*cis*-Androsterone-2,2,3,4,4-d_5 was implemented as the IDS for *cis*-androsterone on October 1, 2011.

[c]Estriol-2,4,16,17-d_4 was implemented as the IDS for estriol on March 17, 2011.

[d]Progesterone-2,3,4-$^{13}C_3$ was implemented as the IDS for progesterone on March 1, 2012.

[e]16-Epiestriol-2,4-d_2 was changed from estriol's IDS to a surrogate compound on March 17, 2011.

[f]Medroxyprogesterone-d_3 was changed from progestrone's IDS to a surrogate compound on March 17, 2011.

solution composition information by USGS personnel for calculation of analyte recoveries in the matrix-spike sample.

7.1.8. *MSTFA injection internal standard (MSTFA/IIS) mixture.* CAUTION: MSTFA is reactive and volatile. It is important that (1) all steps associated with derivatization and extract transfer (section 9.15) are performed in a ventilated fume hood, (2) a hood sash is positioned between the analyst and the extracts, and (3) gloves are worn. See section 3 for safety information.

7.1.8.1. *Cholestane-d_6*: Stock and intermediate solutions (see table 3 for commercial source) are prepared in a non-polar, water-immiscible solvent such as toluene.

7.1.8.2. *Chrysene-d_{12}*: 100 ng/µL or other concentration. Prepared by dilution of a six-component internal standard solution (NSI Solutions Inc., Raleigh, N.C., cat. no. C–394L) into toluene. (The other five components in this mix are not used in this method.) Alternatively, an intermediate standard containing only chrysene-d_{12} can be obtained or prepared from neat material. It is important to ensure that the standard is prepared in a water-immiscible solvent.

7.1.8.3. *N-methyl-N-(trimethylsilyl)-trifluoroacetamide activated II solution*: The MSTFA solution (Sigma-Aldrich Corp. cat. no. 44156–100ML–F) used in this method is supplied from the manufacturer activated with 0.47 percent of 2-(trimethylsilyl)-ethanethiol and 0.18 percent of ammonium iodide (Sigma-Aldrich Company, 2010). Use of non-activated MSTFA solution alone will not achieve adequate derivatization of some method analytes.

7.1.8.4. *MSTFA/IIS mixture*: The MSTFA/IIS mixture is prepared by dilution of cholestane-d_6 and chrysene-d_{12} intermediate standards into 100 mL of the MSTFA activated II solution to yield final concentrations of 2 ng/µL for cholestane-d_6 and 0.1 ng/µL for chrysene-d_{12}. The MSTFA/IIS mixture is stored tightly sealed at <–5°C. A 200-µL aliquot of MSTFA/IIS mixture is used during the derivatization step (see section 9.15).

7.1.9. *Calibration standards*: see section 9.13.

7.1.10. *Third-party check standard*: This multi-analyte standard is used to help verify the performance of the calibration standards and typically is prepared with analyte concentrations that are mid-range of the calibration standards. Analytes used in the third-party check standard are obtained from commercial sources other than those used to prepare the calibration standards. Typically, the third-party check standard is prepared with the steroid analytes at 200 picograms per microliter (pg/µL). Then 100 µL of this mixture is used to prepare the final injection-level third-party check standard (at 100 pg/µL) as described in section 9.13.

8. Sample Collection, Containers, Preservation, Filtration, Shipment, and Holding Times

8.1. Sample Collection

Samples are collected using procedures given in chapter 5 of the USGS National Field Manual (Wilde and others, 2004), with special guidance for these types of samples provided in section 5.6.1.F of that manual (accessed April 2012 at *http://water.usgs.gov/owq/FieldManual/ chapter5/5.6.1.F_v-1.1_4-03.pdf*), except that alternative sample containers are prescribed (see section 8.3). Samples containing potential biohazards require extra caution and special shipping requirements (see section 8.6.2). Some method analytes are found in commonly used products such as prescription drugs. It is important that sampling personnel use care to avoid contamination of the samples by avoiding consumption or contact with such materials immediately before and during sampling procedures. Nitrile or other protective gloves are to be worn when handling samples to minimize risk of personal and sample contamination (see sections 3 and 4).

The method is applicable to either field-filtered (LS 2434) or unfiltered (LS 4434) water samples, and processing of samples by either schedule at the NWQL is procedurally identical. The only difference is with the NWIS parameter and method codes used to distinguish whether the sample was filtered or not. For samples containing high concentrations of particulate matter, it may be appropriate for customers to submit unfiltered water samples for LS 4434 rather than choosing the filtered water option because some analytes will partially or largely sorb to particles.

8.2. Field-Submitted Quality-Assurance Samples

Field-submitted QA samples, including important field blank collection considerations, are described in section 13.4.

8.3. Sample Containers

The following procedures apply to containers used for collection of samples:

8.3.1. The sample size is less than or equal to (≤) 500 mL. Samples that do not contain residual chlorine are contained in 0.5-L high-density polyethylene bottles (USGS National Field Supplies Service, cat. no. Q36FLD) with HDPE-lined screw caps (USGS National Field Supplies Service, cat. no. Q417FLD).

8.3.2. Ascorbic acid is added to samples (50 mg of ascorbic acid per 500 mL of water) collected from sites where halogenation treatment of the water is used (for example, drinking, wastewater, or other facilities using chlorination or bromination disinfection technologies) to quench residual chlorine or bromine, which can react with some of the method compounds (Schenck and others, 2008). HDPE sample bottles containing ascorbic acid are available from USGS National Field Supplies Service (cat. no. N3600).

8.3.3. The container codes to use on the Analytical Service Request form are "HFL" for LS 2434 and "HUN" for LS 4434.

8.3.4. It is important that the sample bottles are not filled completely (sufficient air space is needed for freeze expansion) because samples typically are stored frozen ($\leq$–5°C) before extraction.

8.4. Sample Preservation

Comprehensive sample-preservation studies have not been conducted for these methods, nor are they possible for the wide variety of sample types that might be analyzed. Samples are stored frozen to enhance sample preservation. Ascorbic acid is used for samples that might contain residual chlorine as described in section 8.3.2. Field filtration is preferred for samples intended for LS 2434 to reduce microbial degradation of analytes, with filtering performed as soon as possible following sample collection. Results for holding-time studies for refrigerated and frozen samples of fortified reagent water, along with additional comments regarding sample preservation, are presented in the "Holding-Time Experiments" section.

8.5. Sample Filtration

Samples for LS 2434 preferably are filtered in the field (or at a field laboratory) using the procedures of Sandstrom (1995) as summarized in Wilde and others (2004). Filtration for LS 2434 samples can be performed by the NWQL as indicated by requesting laboratory code 4200 on the Analytical Services Request form.

8.6. Sample Shipment

8.6.1. Sample bottles are shipped overnight to the NWQL on water ice using procedures in NWQL Technical Memorandum 2011.01 (National Water Quality Laboratory, 2011). Samples that can not be sent immediately from field locations are stored in a freezer ($\leq$–5°C) before shipping on water ice.

8.6.2. Samples containing potential biohazards are shipped to the laboratory using procedures given in Wilde and others (2004) and NWQL Technical Memorandum 2011.11 (U.S. Geological Survey, 2011b), and handled using procedures in NWQL SOP MULX0430.x, "Handling of Potential Biohazardous Samples" (Donna Damrau and Helen Wharry, National Water Quality Laboratory, written commun., 2011).

8.7. Sample Storage and Holding Times at the NWQL

8.7.1. Following sample receipt and login at the NWQL, the sample bottle is stored at $\leq$–5°C, unless directed by sample preparation staff to place the bottle in a refrigerator at 4±2°C for those samples for which extraction will commence within 2 working days of receipt, or 3 calendar days if received on a Friday.

8.7.2. Remove frozen samples from the freezer and thaw at ambient temperature (on bench) for about 12 h before extraction, or move to a refrigerator (4±2°C) if thawing the sample over a weekend.

8.7.3. Ideally, sample extraction will commence within 2 working days of receipt, but samples can be stored frozen at $\leq$–5°C for as long as 2 months before extraction, if necessary. Freezing of samples is used for longer-term (days to months) storage for samples that cannot be processed immediately. Freezing reduces biodegradation processes. Results of holding-time studies for refrigerated and frozen samples of fortified reagent water are presented in the "Holding-Time Experiments" section.

8.8. Holding Times for Sample Extracts Before Analysis

8.8.1. Extract storage conditions during sample preparation are described in detail under steps 9.8–9.15 of this report. Ideally, sample preparation is completed with minimal time between procedural steps.

8.8.2. Initiation of sample analysis immediately upon completion of analyte derivatization is essential due to instability of derivatized progesterone (half-life of about 3.5 days at room temperature; substantially longer when held frozen) and is coordinated with the GC/MS/MS analyst.

8.9. Retention and Disposal of Samples and Sample Extracts

8.9.1. Aqueous samples typically are completely consumed during extraction. Replicate samples submitted in HDPE bottles can be stored frozen at $\leq$–5°C until analysis of the primary sample is satisfactorily completed and data are reported, at which time the replicate is disposed.

8.9.2. Following GC/MS/MS analysis, extracts are stored frozen for at least 1 year at $\leq$–5°C along with the co-prepared calibration standards. Under these conditions, extract storage experiments indicate good stability for at least 6 months for all analytes. Furthermore, the use of isotope-dilution quantification increases the robustness of this analysis because evaporative loss of the extract's solvent during extract storage will be compensated for by using IDQ. After at least the minimum retention period, disposal of extracts follows procedures outlined in section 3.

9. Sample Preparation

9.1. Cleaning of General Glassware

Except as noted otherwise in this section, glassware used in this method is cleaned in a dishwasher using Neutrawash® detergent and baked in an oven with a temperature cycle that includes maintaining an upper temperature of 450°C for at least 2 h. All clean glassware is stored in a drawer, or in glass or metal containers with tops, or with the glassware opening wrapped with aluminum foil. These storage conditions reduce contamination from dust. It is important that lab wipes (paper) are not used when handling glassware because paper contains cholesterol (Behrman and Gopalan, 2005); aluminum foil is used instead. Pasteur pipettes and GC vials are baked for at least 2 h at 450°C, and are not silanized. Baked pipettes are stored in metal or glass containers or aluminum foil; the pipettes are not stored in their cardboard boxes.

9.2. Cleaning and Silanization of Specific Glassware

The 50-mL receiver tube, 40-mL SPE receiver tube, and 5-mL reaction vials described in section 5.2 are cleaned, and the interior surfaces are deactivated by silanizing, before use of these specific glassware for each sample preparation event by using the following procedure:

9.2.1. Clean the glassware to be silanized by hand washing (with brush) in hot detergent solution and rinsing well with hot tap water followed by deionized water. Then rinse the glassware with methanol followed by acetone. Visibly soiled glassware might require rinsing with or soaking in an appropriate solvent—methanol, acetone, or dichloromethane—before washing. Bake the glassware for 2 h at 450°C. Allow the glassware to reach room temperature before proceeding to the next step.

9.2.2. Coat the entire interior glassware surface with Sylon-CT™ for at least 1 minute (min). Repeat this coating step two more times. The Sylon CT™ can be reused during each coating step by pouring or pipetting excess Sylon-CT™ from one piece of glassware to the next to complete deactivation of a full set of glassware. CAUTION—Sylon-CT™ is reactive. Wear gloves and perform all work in a ventilated hood (see section 3). Dispose of reagent and methanol rinses as "chlorinated" waste.

9.2.3. Rinse silanized glassware thoroughly with methanol. Following methanol residue evaporation, heat silanized glassware at 100°C for at least 30 min. This glassware is labeled as silanized and stored in drawers, in covered containers, or with the glassware opening covered with aluminum foil.

9.3. Sample Thawing and Set Creation

9.3.1. Retrieve the samples from the freezer or refrigerator; allow frozen samples to thaw overnight at room temperature.

9.3.2. Samples for both LS 2434 and LC 4434 can be processed in the same sample set. The typical set size is 12 samples, including all associated field and laboratory-generated QA and laboratory-specified QC samples, although larger set sizes are permissible.

9.3.3. Set-specific QC samples are added during set creation. Current (2012) laboratory QC sample types available for LS 2434/4434 are: (1) laboratory reagent-water blank (LRB; also called "set blank" or "PBLNK") that has an 8000-series NWQL sample identification number; and (2) laboratory reagent-water spike (LRS; also called "set spike" or "PSPK") that has a 9000-series NWQL sample identification number. Additional laboratory-selected QA samples might include laboratory matrix duplicate samples (DUP) and laboratory matrix spike samples (MSPK). Additional descriptions are provided in section 13.

9.4. Extractor Cleaning

9.4.1. Wash the stainless-steel extractor tubes; end caps; frits; and the Teflon®, Viton®, and silicon O-rings with hot Liquinox® detergent solution using tube brushes. Rinse all of these parts thoroughly with hot tap water, then deionized water; allow the parts to drain for at least 10 min.

9.4.2. Rinse the Viton® and silicon O-rings with methanol and set them on aluminum foil to dry. Rinse the SSE tubes and caps with methanol and allow to dry. Then rinse the tubes and caps with acetone and allow to dry. Note: Viton® O-rings are not to be rinsed with acetone. If additional solvent rinsing is required, hexane is used.

9.4.3. Sonicate the frits and Teflon® O-rings in a beaker for at least 30 min in methanol. Pour off the methanol, and rinse the frits and O-rings again with methanol followed by acetone. Place the frits and O-rings on acetone-rinsed aluminum foil in a fume hood to dry briefly. Store the frits and O-rings in a sealed container.

9.5. Extractor Assembly (Pre-cleaned)

9.5.1. Clamp the clean SSE tubes onto the manifold panel rack and tighten the clamps with the wing nuts (fig. 3). Place a silicon O-ring in the top tube groove and Viton™ O-ring in the bottom tube groove. Wrap the bottom thread of the extractor tube with Teflon tape; use of this tape makes cap sealing and unsealing easier.

9.5.2. Using new, clean nitrile gloves, place a stainless steel frit in the bottom end cap. Using filter forceps, put one ENVI™ C_{18} disk (wrinkle side up) on top of the frit. It is important to ensure that the disk is positioned correctly within the cutout of the bottom end cap (fig. 4A; note: the frit and C_{18}

disk are not shown in figure 4*A* and, although positioned in the bottom cap, are not visible in figure 4*B*).

9.5.3. Place one glass-fiber filter (fuzzy side up) on top of the ENVI™ disk. Ensure that the GFF is positioned correctly within the cutout of the end cap (fig. 4*B*).

9.5.4. Wet the GFF and C$_{18}$ disk with sufficient methanol to facilitate seating of the Teflon® back-up O-ring. Press-fit the Teflon® O-ring on top of the GFF. Ensure that the outer edge of the O-ring seals snugly within the cutout of the end cap by pressing down around the edge of the O-ring with forceps or gloved fingers (fig. 4*B*). Carefully screw the bottom end cap onto the bottom of the tube and ensure that the seal is tight.

9.6. Quality-Control Sample Preparation and Isotope-Dilution Standard Addition

9.6.1. Dispense approximately 450 mL of reagent water into two HDPE sample bottles. Label one bottle with the LRB (or set blank) NWQL sample identification number and the other with the LRS (or set spike) NWQL sample identification number.

9.6.2. Weigh the sample in the bottle to ±1 g, and record this combined weight of the sample (*S1*) plus bottle with cap (*B*) for all field and QC samples in the set. These weights are used in the equation presented in section 11.1.

9.6.3. Use the scoopula to add about 50 mg of NaCl (salt) to all samples.

Note: The sterols 3β-coprostanol and cholesterol (and the IDS cholesterol-*d$_7$*) had poor recovery in the reagent-water matrix only (containing no salt) because of incomplete isolation on the C$_{18}$ disk. NaCl is added to all samples, but the primary purpose for its addition is to enhance recovery of these compounds in the various reagent-water matrices (see "Cholesterol-*d$_7$* Recoveries in Non-salted Reagent Water" section).

9.6.4. Using the IDS 100-µL microdispenser, add 100-µL of the isotope-dilution standard fortification mixture (IDS mixture; see section 7.1.4) to all samples in the set. Cap the sample bottle and shake the bottle to mix the sample.

9.6.5. Using only the spike 100-µL microdispenser, add 100-µL of the 2434/4434 spike mixture (see section 7.1.7) to the LRS (set spike; PSPK), to the laboratory matrix spike sample (MSPK; if used), and to the field-requested laboratory matrix-spike sample (FRLMS; if requested). Note: The volume of 2434/4434 spike mixture added to a given MSPK or FRLMS might be greater than 100 µL based on anticipated ambient concentrations of analytes in the sample matrix.

9.7. Sample Extraction

The following procedures are used for sample extraction (refer to fig. 3):

9.7.1. Place a plastic waste beaker beneath each SSE to catch solvent rinses and extracted sample.

9.7.2. Rinse the interior of the SSE tube with a total of 40 mL of methanol using bottle-top dispenser. The four 10-mL portions of methanol are directed alternately down opposite sides of SSE tube to achieve rinsing of the entire surface of the tube wall.

9.7.3. Allow the methanol rinse to elute from the tube without applying N$_2$ for at least the first 10 mL of rinse. Observe the flow of methanol from the tube and take the following actions as needed:

9.7.3.1. If the flow is unusually slow in a tube compared to the other tubes and to normal expected flow (which is a slow stream or steady drip), then attach the top cap and apply minimal N$_2$ pressure of ≤14 kilopascals (kPa)). These conditions usually allow good flow to be reestablished. If not, and more pressure is required to achieve typical flow, then force the remainder of the methanol to exit the tube by applying N$_2$ higher pressure. Relieve the N$_2$ pressure to the SSE by turning the 3-way valve handle up. Unscrew the bottom cap from the tube and remove the C$_{18}$ disk, GFF, and Teflon® O-ring from the cap. Place a new C$_{18}$ disk, GFF, and O-ring in the cap as described in sections 9.5.2 to 9.5.4, and then retest the SSE beginning at step 9.7.2.

9.7.3.2. If non-pressurized flow is too fast compared to other tubes and to normal flow, then the following are performed to rectify:

9.7.3.2.1. Verify that the bottom cap is tightly sealed on the SSE tube. If not, then tighten the cap, and check that the solvent flow has returned to normal. If this step corrected the flow, then proceed to step 9.7.4.

9.7.3.2.2. Disassemble the bottom cap and ensure that both the C$_{18}$ disk and GFF were installed. If not, then add the missing component, reseal the cap to the tube, and retest the SSE beginning at step 9.7.2.

9.7.3.2.3. If both the disk and GFF were installed correctly, then replace the C$_{18}$ disk, GFF, and Teflon® O-ring and retest the SSE beginning at step 9.7.2.

9.7.4. Following complete elution of the methanol rinse, add 25-mL reagent water to the SSE tube and allow the water rinse to drain using low applied N$_2$ pressure if needed (≤14 kPa).

Note: the C$_{18}$ disk is kept moist and "activated" by ensuring that the water rinse is added within 10 min of complete methanol elution, and that sample addition (section 9.7.5) likewise occurs within 10 min of complete elution of the water rinse. It is important that N$_2$ is not applied to the SSE at the conclusion of either of these elution rinse steps, because it might cause the disk to begin to dry.

9.7.5. Shake the sample bottle vigorously, and then carefully pour the sample into the extractor tube.

Notes: For samples in bottles with capacities greater than (>) 500 mL, it is important to ensure that the water level does not exceed tube volume. For these, the sample water is added to the tube in smaller aliquots. For some samples, it may be necessary to increase the flow rate by applying N$_2$ pressure to the SSE before the entire sample has been added (see section 9.7.7). If the SSE must be opened during the extraction process, the SSE is vented to ambient pressure before opening.

9.7.6. After the entire sample has been poured into the SSE, add 25-mL of reagent water to the sample bottle. Cap the bottle, shake the bottle well, and then pour the water rinse from the bottle into the SSE tube.

9.7.7. Place the top end cap on the SSE (to minimize airborne dust contamination potential), and set the 3-way valve to the vent (up) position. Allow the sample to elute through the GFF/C_{18} disk under its own static pressure for at least 30 min. If the flow becomes slow, check that the top cap is screwed completely onto the SSE, and slowly apply N_2 pressure to the tube to speed up extraction. Regardless of the amount of N_2 pressure applied, it is important that the sample flow exiting the drain tube never exceeds a steady, rapid drip rate where distinct droplets (but not a stream flow) are visible. Increasing pressure to specific SSEs might be required during the extraction process because sample particles load onto the GFF/C_{18} disk and retard flow. Pressures required typically are in the 14–34 kPa range and rarely exceed 138 kPa. The SSE can be pressurized to as much as 790 kPa and is limited to this maximum by the gas regulator used. Pressures exceeding 480 kPa are avoided if possible. If exceeded, the hood sash is positioned between the sample preparation personnel and the tube(s) at higher pressure while any adjustments are being made, and the sash is completely closed when adjustments are not in process. Samples that clog (so that no flow goes through GFF/C_{18} disk) or flow very slowly at pressures nearing 790 kPa likely can not be completely extracted within a reasonable time (about 4 hours). In this case, the procedure described in the "Note 3: Unextracted Sample Portion" of this section is followed.

Note 1. Before applying N_2 pressure to the SSE, ensure that the top cap is screwed onto the SSE so that it seals against the silicone O-ring. As pressure is increased using the 2-stage gas regulator, it is important to ensure that the sample flow for all pressurized samples (those with the 3-way valve in the down [pressurized] position) does not exceed a rate where distinct droplets still occur (not a stream flow). The 3-way and needle valves are used to assist with pressure application and to maintain desired flow. Once tube pressurization during sample extraction exceeds 34 kPa, the extraction process is continuously attended and monitored.

Note 2. Once pressurized, pressure inside the SSE is relieved by turning the 3-way valve handle to the up (vent) position. The N_2 is vented from the SSE tube before the top cap is removed.

Note 3. Unextracted Sample Portion—Samples with large amounts of fine (clay, colloid) particles might clog the GFF/C_{18} disk resulting in minimal or no subsequent sample flow even at high N_2 pressures (550–690 kPa) following partial sample extraction. In this case, only a portion of the sample will be extracted through the GFF/C_{18} disk. If this occurs, the remaining unextracted portion of the sample is poured back into the sample bottle using a clean funnel placed in the bottle opening. The bottle (with cap) plus the unextracted sample ($B+S_u$) is weighed, and this weight is recorded (see section 9.7.10). Of the wide variety of samples processed during

method validation and custom implementation, GFF/C_{18} disk clogging occurred for only one unusual sample (a high-solids manure-water slurry).

Proceed to the next step once certain that sample extraction flow is easily maintained and will not clog the GFF/C_{18} disk (usually after about one-half the sample has been extracted).

9.7.8. While the sample is extracting, add 10 mL of the 25-percent methanol/water solution to the empty sample bottle. Cap and shake the bottle vigorously to coat the entire inside surface with the rinse.

9.7.9. After the entire sample has eluted through the GFF/C_{18} disk, shake the sample bottle again and pour the 25-percent methanol/water rinse from the sample bottle into the extractor tube. Allow this rise to pass through the GFF/C_{18} disk at the typical drip rate; slowly apply N_2 pressure if needed. For each sample, the pressure required to force the 25-percent methanol/water solution through the disk usually is 7–35 kPa; however, if more pressure is required, the pressure should be less than the final pressure required during extraction.

9.7.10. Weigh the empty bottle with cap (B) on the balance to ±0.1 g, and record this weight. Note: If the entire sample could not be extracted (see Note 3 in section 9.7.7), after weighing the bottle (with cap) containing any unextracted sample [$B+S_u$] as described in Note 3 in section 9.7.7, discard the residual sample in the bottle, and weigh the empty bottle with cap (B). These weights are used in the equation presented in section 11.1.

9.7.11. Dry the GFF/C_{18} disk by applying N_2 pressure at 140 kPa for at least 10 min., but no more than 20 min. If the final pressure during extraction was greater than 140 kPa, this step is carried out at the final extraction pressure. Note: the subsequent extract evaporation step (section 9.9) might be slowed substantially by the presence of residual water in the SSE that ends up in the methanol disk eluent (section 9.8).

9.7.12. Rinse the exterior of the bottom end cap's drain tube with about 10 mL of methanol.

9.7.13. Discard the extracted water and solvent rinses contained in the the 1-L plastic beaker as nonchlorinated hazardous waste.

9.8. Elution of GFF/C_{18} disk

9.8.1. Position a 50-mL receiver tube that is labeled with the sample identification number (held in rack or in beaker) under the SSE drain tube. Note: This glassware was cleaned and silanized before use as described in section 9.2.

9.8.2. Dispense two 20-mL portions of methanol into the SSE tubes. Add each 20-mL portion down opposite sides of SSE tube to achieve rinsing of the entire interior tube wall surface. Collect the eluent into the silanized 50-mL receiver tube. Place the top end cap on the tube but do not tighten the cap. Allow the elution to proceed unpressurized for 30 min, and then tighten the cap and apply N_2 pressure to 138 kPa, or greater as needed, to elute the remaining methanol.

9.8.3. Remove the receiver tube and proceed immediately to the first solvent evaporation step (section 9.9), or cap the receiver tube with a clean 24/40 stopper and store the extract for as long as 2 working days at 6°C or lower.

9.9. First Evaporation

9.9.1. Fill the evaporator bath with distilled water to a depth of about 5 cm. Set the thermostat for the bath to yield a water temperature of 25±2°C. This setting will maintain the extract temperature near 25°C as evaporative cooling occurs. Verify the bath temperature with a thermometer before starting evaporation.

9.9.2. Place the 50-mL receiver tubes on the N-Evap (section 5.9.1), and attach the needles to the evaporator. Evaporate the extract to dryness under a gentle N_2 stream using the following initial settings:

- Immerse the bottom of the tube into the water bath. Open the needle valves on all positions before applying N_2 flow.

- Set the N-Evap flow meter to 2–5 L per min based on the number of samples requiring evaporation.

- Adjust the N_2 flow for each sample using the needle valves to yield a dimple indentation in the solvent surface of 0.75–1 cm; use a flow that minimizes excessive splashing.

- Adjust the needle position downward into the tube as the solvent level lowers to facilitate timely evaporation.

This evaporation step usually is performed overnight, but do not exceed 19 h unmonitored (see following note). Samples with more residual water might take longer than 19 h to dry.

Note: For samples with higher amounts of dissolved organic matter, a residue will form and remain on the tube bottom after the solvent is dry. This residue might look "wet," but evaporation is terminated if the volume or appearance of this residue does not change after an additional 45 min of maximum evaporation time since last monitored and looking "nearly dry." To test whether residual water might be present, add about 1 mL of DCM to the tube, cap the tube, and mix the solution using a vortex mixer. If two immiscible phases form, water is present; return the tube to the N-Evap, and evaporate the DCM and residual water to dryness. If only one phase is present, then water likely is not present; return the tube to the N-Evap and evaporate to dryness. Once the DCM has evaporated, evaporation is terminated if the final volume or appearance of the final residue does not change after an additional 30 min of maximum evaporation time since looking "nearly dry."

9.9.3. Remove the receiver tube from the N-Evap and cap the tube with a 24/40 stopper. Proceed immediately to section 9.10: "Florisil Cleanup of Extract," or store the extract for as long as 2 working days at 6°C or lower.

9.10. Florisil Cleanup of Extract

The following procedures are used for the Florisil cleanup of the extract (refer to fig. 5):

9.10.1. Add 2 mL of 5-percent methanol/DCM to each sample. Mix the sample using the vortex mixer for 5–10 seconds, and then set the sample aside for at least 30 min before introduction to the Florisil column. Note: Enhanced recoveries of certain analytes are obtained when sufficient time is allowed for dissolution of the dried extract. It is important that this 30-min time is not shortened.

9.10.2. Open the column valves for the six designated positions on the top of the Visiprep™ vacuum manifold; the other column valve positions remain closed. Rinse the six open valves with acetone and insert new valve liners into these six valves.

9.10.3. *Florisil column precleaning*: Attach the Florisil SPE column to the Luer hub of the valve liner. Add five 5-mL aliquots of acetone to clean and activate the column. Allow each aliquot to elute by gravity flow. After all 25 mL have passed through the column, dry the column for at least 5 min (but for no more than 10 min) under vacuum pressure (about 13 kPa).

Rinse the Florisil column with two 5-mL portions of 5-percent methanol/DCM. Allow the solvent to drip through by gravity; do not apply vacuum before moving to the next step.

9.10.4. Remove the top of the Visiprep™ vacuum manifold (with columns attached) from the chamber. Close the chamber's vacuum control valve, turn on the vacuum, and tilt the chamber towards the drain (front) to aspirate the rinse solvent into the waste carboy. Turn off the vacuum.

9.10.5. Insert six silanized SPE receiver tubes (labeled with sample identification numbers) into the Visiprep™ collection rack. Replace the Visiprep™ top on the chamber. Note: Ensure that all valve liners are inside the SPE receiver tubes.

9.10.6. Carefully transfer each sample extract to the Florisil column by using a Pasteur pipette.

9.10.7. Dispense another 2-mL rinse aliquot of 5-percent methanol/DCM to each sample's 50-mL receiver tube. Mix this aliquot in the receiver using the vortex mixer. After the initial 2-mL extract portion passes into the sorbent bed, transfer the 2-mL rinse aliquot from the receiver tube to the column by using the pipette.

Note: Six samples can be processed one at a time through steps 9.10.6–7, or all the initial transfers (9.10.6) can be completed for all six samples in sequence, followed by completion of the tube-rinse step (9.10.7) for all six samples in sequence.

9.10.8. After the 2-mL rinse aliquot passes into the sorbent bed, add four 5-mL portions of 5-percent methanol/DCM to the column using the dispenser. Maintain the solvent elution flow rate at a slow (unimpeded gravity) drip rate.

9.10.9. If the flow rate for a column is too slow, then vacuum may be applied to maintain the desirable flow rate. However, apply any vacuum carefully so that it does not dramatically increase the flow for those columns having unimpeded flow.

9.10.10. After the final 20 mL (the four 5-mL portions) of the 5-percent methanol/DCM per column has passed through the 6 columns (assuming all 6 positions were used), apply the vacuum at a pressure of about 14 kPa to draw the residual solvent into the SPE receivers.

9.10.11. Turn off the vacuum, discard the pipettes and columns, and remove the Visiprep™ top. Carefully lift out each SPE receiver tube and proceed immediately to the second evaporation step. If the second evaporation step will not immediately follow this step, then cap the SPE receiver tube with a 19/22 stopper and store the extract for as long as 2 working days at 6°C or lower.

9.10.12. Repeat steps 9.10.2 through 9.10.11 for the remaining samples in the set.

9.11. Second Evaporation

9.11.1. Place the SPE receiver tubes on the N-Evap evaporator and attach the needles. The tubes are not immersed into the water bath because the bath is not needed for the second evaporation step. Concentrate the extract to 1–2 mL (over approximately 2 hours) with N_2 (see section 9.9 "First Evaporation" for appropriate gas flow rates and instructions on how to load the N-Evap). Remove the SPE receiver from the evaporator, cap the receiver with a 19/22 stopper, and mix the extract in the receiver using the vortex mixer.

9.11.2. Proceed to the next step or, if needed, store the extract in the capped SPE receiver for as long as 2 working days at 6°C or lower.

9.12. Transfer Extract to Reaction Vial

9.12.1. Carefully transfer the extract to a silanized 5-mL reaction vial (which is labeled with the sample identification number and held in vial rack) using a Pasteur pipette.

9.12.2. Using the dispenser, add 1.5 mL of 5-percent methanol/DCM to the SPE receiver. Mix this rinse aliquot in the receiver using the vortex mixer, and transfer the rinse aliquot to the reaction vial. Repeat this receiver rinse step.

9.12.3. Proceed immediately to the third evaporation step (section 9.14) or, if needed, cap the reaction vial and store the vials for as long as 2 working days at 6°C or lower. Note: ensure that the septum is fully inserted into the cap before screwing the cap onto the reaction vial.

9.13. Preparation of Calibration Standards

Method compounds in instrument-calibration standards also must be derivatized before GC/MS/MS. Due to the limited stability of the derivatized form of progesterone at room temperature, calibration standards older than 3 days are not used unless the purpose is to rerun a set of samples of similar age to the standards. Therefore, calibration standards

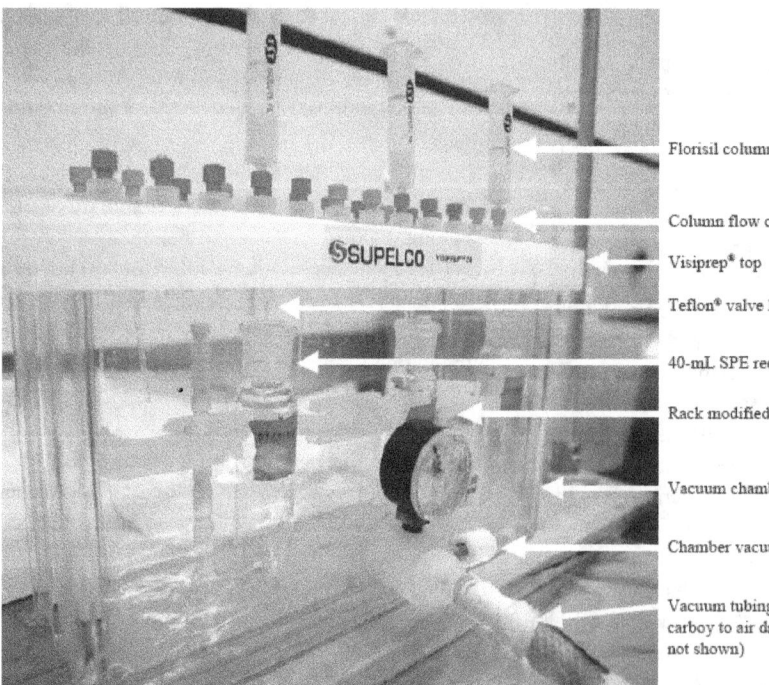

Florisil column

Column flow control valve

Visiprep® top

Teflon® valve liner

40-mL SPE receiver tube

Rack modified to hold 40-mL tube

Vacuum chamber

Chamber vacuum control valve

Vacuum tubing (connects via waste carboy to air driven vacuum pump; not shown)

Figure 5.   Vacuum manifold used for extract cleanup with Florisil solid-phase extraction (SPE) columns (mL, milliliter).

are prepared more frequently for the hormone methods than is routine for other NWQL methods. The standards are prepared as a part of the sample preparation set, or, more typically, with one of several sample preparation sets when multiple preparation sets are grouped into an instrumental batch. As such, after the reaction vials containing the calibration standards are prepared using the following procedures, these calibration standard vials are further processed along with the sample extracts of a given sample preparation set beginning with step 9.14.

9.13.1. Remove the high (section 7.1.5) and low (section 7.1.6) analyte mixtures, the third-party check standard (section 7.1.10), and the IDS mixture (section 7.1.4) solutions from the freezer and allow to warm to room temperature, which typically takes less than 2 h.

9.13.2. Label ten 5-mL conical reaction vials that have been cleaned and silanized (per section 9.2) with the instrument standard names shown in table 4. Note: More than 10 vials are needed if additional calibration or verification standards are being prepared; for example, higher level calibration standards.

9.13.3. Add about 1–2 mL of DCM to each reaction vial before addition of IDS or analyte standard solutions.

9.13.4 Add 100 µL of the IDS mixture to each of the vials using the IDS 100-µL micro dispenser (see section 5.6). Note: It is important to ensure that the IDS solution lot number is the same as that of the solution used to fortify the samples using the procedures in section 9.6, and to ensure that the IDS is added to the vials before any analyte additions.

9.13.5. Add the appropriate volume of either the 0.01-ng/µL (low) or the 1-ng/µL (high) analytes mixture, or the third-party check standard to the labeled standard vials (containing DCM) as shown in table 4. Use the 25-µL syringe for adding the 10 and 20-µL volumes. Use the 100-µL syringe for adding the 50, 100, and 200-µL volumes.

Note: The 1CAL–10CAL calibration standards are prepared first, then the 50CAL–1000CAL calibration standards are prepared. The syringes are rinsed well with methanol and acetone before and after calibration preparation. It is important to ensure that no residual solvent remains in the syringe barrel before starting additions.

9.13.6. Proceed immediately to the third evaporation step or, if needed, cap the calibration standard reaction vials and store for 1 working day at 6°C or lower. Note: ensure that the septum is fully inserted into the cap before screwing the cap onto the vial.

9.14. Third Evaporation

9.14.1. Adjust the bottom plate on the N-Evap to a position that will hold reaction vials in place. Ensure that the set screws holding the plate are tight. Carefully place the reaction vials on the N-Evap. Attach the needles to the N-evap, and position

Table 4. Preparation scheme for instrument calibration (CAL) and third-party check (TPC) standards.[a]

[IDS, isotope-dilution standard; IIS, injection internal standard; LIMS, laboratory information management system; MSTFA, *N*-methyl-*N*-trimethylsilyl trifluoroacetamide; ng/µL, nanograms per microliter; pg/µL, picograms per microliter; µL, microliter]

Instrument standard name (LIMS name)	Desired final concentration of analytes[b] (pg/µL)	Volume of IDS mixture at 0.5 ng/µL (7.1.4)[c] (µL)	Volume of low analytes mixture at 10 pg/µL (7.1.6) (µL)	Volume of high analytes mixture at 1,000 pg/µL (7.1.5) (µL)	Volume of third-party check standard at 200 pg/µL (7.1.10) (µL)	Volume of MSTFA/IIS mixture (7.1.8) (µL)
0CAL	0	100	0	0	0	200
1CAL	1	100	20	0	0	200
5CAL[d]	5	100	100	0	0	200
10CAL	10	100	200	0	0	200
50CAL[e]	50	100	0	10	0	200
100CAL[f]	100	100	0	20	0	200
250CAL	250	100	0	50	0	200
500CAL	500	100	0	100	0	200
1000CAL	1,000	100	0	200	0	200
TPC[g]	100	100	0	0	100	200

[a]Preparation of the calibration and TPC standards for instrumental analysis is described in section 9.13. Higher levels of calibration standard can be used to extend the calibration range at the analyst's discretion based on anticipated analyte concentration in the samples processed in the sample preparation set.

[b]Concentrations of cholesterol and coprostanol are 100 times higher, and bisphenol A is 10 times higher, than the other method analytes in the analyte mixtures.

[c]Section of report where standard mixture is described.

[d]The 5CAL standard is reanalyzed as the instrumental detection level standard (5IDL in LIMS) within an instrument-analysis sequence (see table 8).

[e]The 50CAL standard is reanalyzed as the continuation calibration verification standard (50CCV in LIMS) within an instrument-analysis sequence (see table 8).

[f]If needed, the 100CAL standard is reanalyzed as the 100CCV within an instrument analysis sequence.

[g]The TPC standard is prepared using all or a subset of analytes obtained from alternative commercial sources than those used for preparation of the analyte mixes. Other concentrations can be used for the TPC as needed.

the outlet end of the needle to be at or just above the tops of the vials. The vials are not immersed into the water bath because a water bath is not needed for this step.

9.14.2. Open the valve of one unused needle position (without the needle attached). Valves for all positions with vials are then opened minimally. Slowly turn on the main N_2 gas flow, and adjust the N-Evap flow meter to give 2–5 L per min gas flow based on the number of samples. The N_2 flow is adjusted as needed by manipulation of the needle valves for each sample position and for the unused position to give a dimple indentation in the solvent surface of <0.2 cm.

9.14.3. Evaporate the extracts to dryness (about 1 h) with N_2 gas. The extracts are immediately removed upon dryness.

9.14.4. Cap the reaction vials. If the GC/MS/MS is ready for samples (see following Note), then proceed to the derivatization step (section 9.15), or store the vials overnight at 4±2°C until commencement of step 9.15 the next day. If the GC/MS/MS is not ready, the reaction vials can be stored for as long as 50 working days (or longer if necessary) at –5°C until notified by the GC/MS/MS analyst to proceed to step 9.15.

Note on extract processing and storage before GC/MS/MS: It is important that the GC/MS/MS analysis commences immediately following the derivatization step 9.15. Therefore, sample-preparation staff will carefully coordinate with the GC/MS/MS analyst on readiness of the instrument before proceeding to step 9.15. If the GC/MS/MS is not ready, extracts are best processed through step 9.14.4 and then stored frozen until the method analyst says to proceed to step 9.15. It is best if the storage time before derivatization is not more than 8 weeks. However, if GC/MS/MS analysis can not be performed within 8 weeks because of prolonged instrument in-operation or analysis backlog, then extended storage as dry extract (at step 9.14.4) is preferred over extended storage of derivatized extracts.

9.15. Derivatization and Extract Transfer to GC Vial

Caution: MSTFA is reactive and volatile. All of the following procedures are performed in a fume hood, with the hood sash positioned between the sample preparation staff and the extracts. Gloves are worn. See section 3 for detailed safety information.

9.15.1. Turn on the heating block heater and set the thermostat to give a block temperature of 65°C; verify that the block has reached this setpoint temperature with a bulb thermometer.

9.15.2. Remove the MSTFA/IIS mixture (section 7.1.8) from the freezer and allow the solution to warm to room temperature (about 2 h). Remove any stored sample extacts and calibration standards contained in reaction vials from the refrigerator or freezer and allow to warm to room temperature.

9.15.3. Using the MSTFA/IIS 200-μL microdispenser with glass bore, add 200-μL of the MSTFA/IIS mixture to each field and QC sample reaction vial, and to each calibration standard reaction vial. If multiple sets (more than 48 vials) are

being prepared, only samples that can be immediately put on the heating block are derivatized; excess samples are held until derivatization of the first set is complete.

9.15.4. Cap each reaction vial (ensure that the septum is properly seated in the cap before screwing the cap onto the vial). Mix the extract using the vortex mixer for 5–10 seconds. Place the reaction vial in the heating block at 65°C for 60 min.

9.15.5. Remove the vials from the block and allow the vials to cool to room temperature (about 1 h). If additional samples remain to be derivatized, they are processed beginning at step 9.15.3.

9.15.6. Using a Pasteur pipette, carefully transfer each extract to a GC vial with low-volume insert and labeled with the sample's identification number. Cap each vial and proceed immediately to the GC/MS/MS analysis (section 10). Note: The derivatization step is not performed unless the analyst has prepared the GC/MS/MS to undertake immediate analysis. If there are circumstances that prevent immediate analysis, then the GC vials are placed in a freezer at –5°C or lower.

10. Analysis by Gas Chromatography with Tandem Mass Spectrometry

10.1. Overview

Sample extracts ready for instrumental analysis by GC/MS/MS contain isotope-dilution standards and injection internal standards in addition to any method analytes that were present in the original sample. The extract solvent is the derivatization reagent, N-methyl-N-trimethylsilyl trifluoroacetamide. Derivatization increases molecular weight and volatility of the target compounds while protecting polar functional groups, making them more amenable to separation by capillary-column GC.

During derivatization with MSTFA, all method compounds lose at least one hydrogen atom, either from any alcohol (C–OH) function groups or from an adjacent (alpha) carbon to any carbonyl (C=O) functional groups, or from both functional group types if present on the same molecule. When carbonyl functional groups on the method compounds are derivatized, the C=O double bond shifts into the adjacent alkane or alkene chain or ring as the trimethylsilyl-enol-ether derivative is formed. The resulting change in valence displaces a hydrogen atom, as is shown for testosterone in figure 6. Bisphenol A-d_{16} loses two phenolic deuterium atoms during derivatization. The IDS name bisphenol A-d_{16} (table 2) used for data reporting in NWIS is the form spiked into samples and, thus, before deuterium loss during derivatization.

After GC separation, the method compounds are introduced to a Quattro-micro-GC™ (Waters Corp.; section 5.17.2) for tandem mass spectrometric analysis where the method compounds undergo electron-impact ionization followed by multiple-reaction monitoring (MRM) of three unique precursor-to-product transitions as described in detail in section 10.3. Subsequent data analysis is performed using TargetLynx™ software (Waters Corp.; section 5.17.4), and a program written at the NWQL is used to validate and format data for export to the laboratory information management system before its export to NWIS.

10.2. Separation by Gas Chromatography

After sample extraction and derivatization, 2 μL of each 200-μL extract is injected on an Agilent 6890 GC equipped with a 7863B autosampler (Agilent Technologies, section 5.17.1). The injection system is operated in splitless mode at 275°C with a helium (section 6.17.1) carrier-gas-flow rate of 1 milliliter per minute and a 1-min hold time before the injection port liner is purged. A Siltek®-deactivated injection-port liner (section 6.17.3) is used to minimize compound decay in the injection system. Method compounds are then separated on a Restek Rxi®-XLB capillary column (section 6.17.4) using a 7-step program to control oven temperature (table 5). The derivatized hormones are structurally very similar; they share a common polycyclic backbone that differs only by substituent attachment and degree of saturation. As a result, the derivatized hormones are relatively difficult to separate chromatographically and a slow change in GC oven temperature is necessary in the range where most compounds elute. A typical separation of the method analytes only in a calibration standard is shown in the gas chromatogram in figure 7 (the IDS and IIS compounds are not shown). Cholesterol and 3β-coprostanol elute several minutes after the other 18 method analytes, and are omitted from figure 7 to maintain scale. The interface between the GC oven and the evacuated mass-spectrometer source is maintained at 300°C. After each chromatographic cycle, the GC oven is cooled to 100°C and allowed to equilibrate for 3.5 min before the initiation of the next injection.

10.3. Tandem Mass-Spectrometry (MS/MS) Analysis

After GC separation, the column effluent is ionized by electron impact at 70 electron volts with the ion source temperature maintained at 230°C. For most method analytes and IDS compounds, the molecular ion (M+) has the highest relative abundance in the resulting mass spectrum and is selected as the precursor ion for MRM analysis. For a few compounds, a fragment ion, usually M+–15 or M+–90, has substantially higher abundance and is selected as the precursor ion. The charged ions are forced out of the ion source by a repeller voltage, and the resulting ion beam is focused by a series of lenses to the mass analyzer. The first quadrupole is used to select for the chosen precursor ion and filter out unwanted ions. This precursor ion is passed into a hexapole collision cell pressurized with a maximum pressure of 0.4 pascals (Pa) with argon (section 6.17.2) and fragmented by collision-induced dissociation. The collision-induced-dissociation voltage is optimized for each individual MRM transition. After collision-induced dissociation, the fragment ions pass through a second quadrupole where the desired product ions are selected and unwanted ions are filtered out.

Because of the added specificity inherent in MRM analysis, mass resolution for the quadrupoles is set fairly wide to maximize throughput of the target ions. For example, the resolution setting gives a peak width at one-half height of about 0.4 atomic mass unit for the 414-m/z ion of perfluorotributylamine used for tuning. The uniquely identified ions from selected MRM transitions are finally directed to an off-axis detector equipped with a photomultiplier for signal enhancement. Example GC retention times (using the specific analysis shown for the analytes only in fig. 7), along with the precursor and product ions, and the collision-induced-dissociation voltage for each MRM transition for all the method compounds, are shown in table 6.

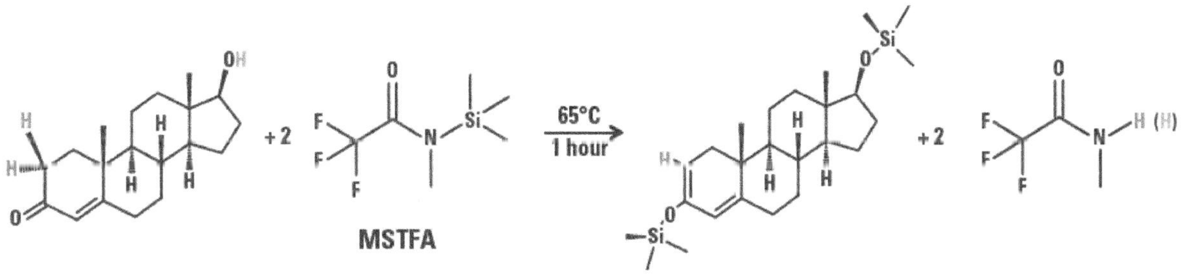

Figure 6.   Derivatization of testosterone with *N*-methyl-*N*-trimethylsilyl trifluoroacetamide (MSTFA) to form the di-(trimethylsilyl)-testosterone derivative. Loss of either of the two alpha hydrogen atoms (shown in red) occurs from the (alpha) carbon adjacent to the ketone functionality to form the enol-ether derivative.

Table 5. Oven temperature program used for analysis of method compounds by gas chromatography (GC) with tandem mass spectrometry.

[C, degrees Celsius]

GC oven ramp number	Rate (°C per minute)	Ramp end temperatue (°C)	Ramp end hold time (minute)
Initial temperature	0	100	1
1	25	235	0
2	1	240	5
3	1	245	0
4	2	265	0
5	10	275	2
6	40	310	5

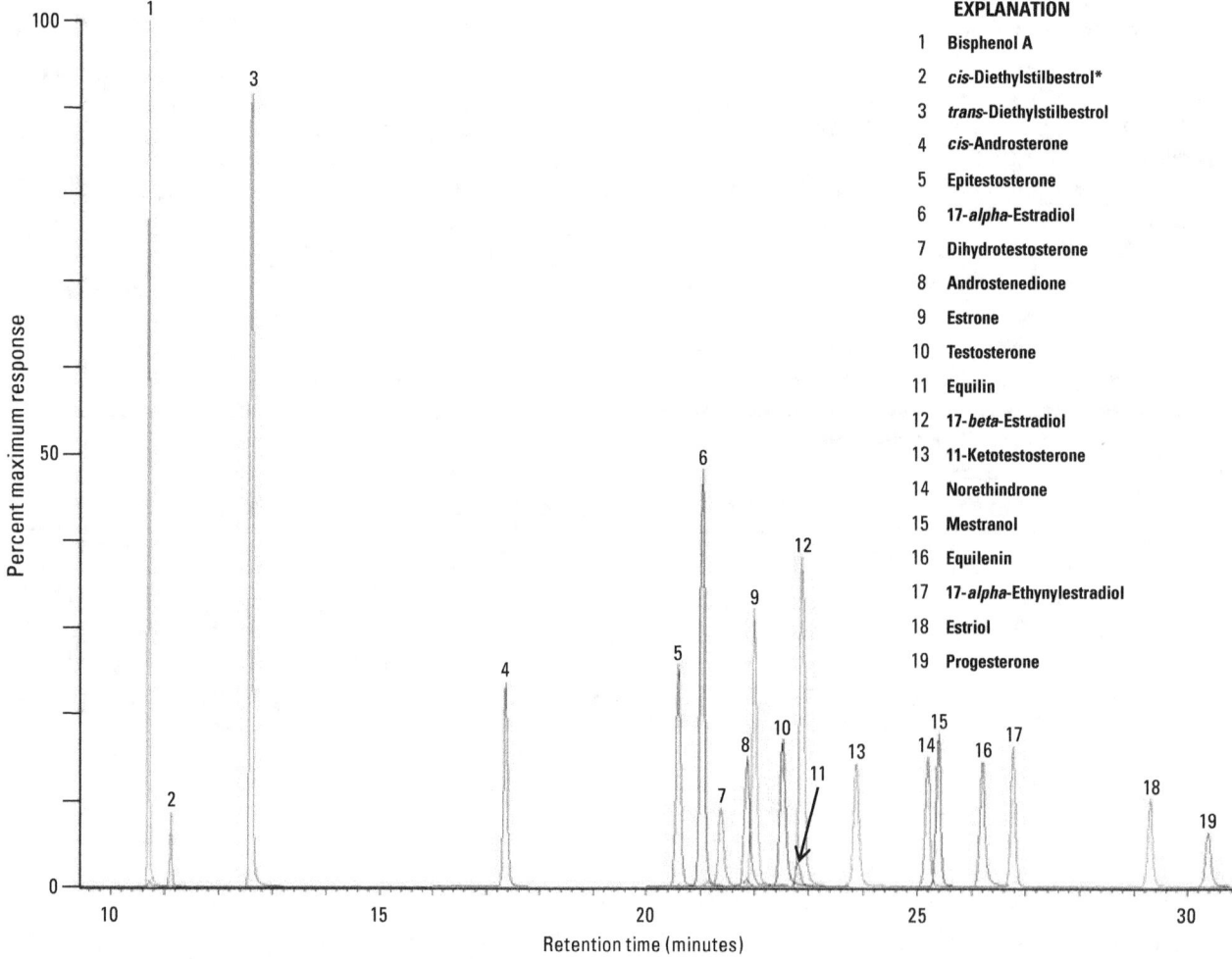

EXPLANATION

1 Bisphenol A
2 *cis*-Diethylstilbestrol*
3 *trans*-Diethylstilbestrol
4 *cis*-Androsterone
5 Epitestosterone
6 17-*alpha*-Estradiol
7 Dihydrotestosterone
8 Androstenedione
9 Estrone
10 Testosterone
11 Equilin
12 17-*beta*-Estradiol
13 11-Ketotestosterone
14 Norethindrone
15 Mestranol
16 Equilenin
17 17-*alpha*-Ethynylestradiol
18 Estriol
19 Progesterone

Figure 7. Example gas chromatrogram showing separation of 18 method analytes. The analytes 3-*beta*-coprostanol and cholesterol elute at later retention times and are not shown. Isotope-dilution, surrogate, and injection internal standard compounds are not plotted. *The *cis*-diethylstilbestrol isomer is not determined by the method.

Table 6. Parameters for multiple-reaction monitoring (MRM) analysis of derivatized method compounds and other compounds used in the tandem mass-spectrometry acquisition method.

[The precursor ion (**boldface** value) is the same for each MRM transition (except 11-ketotestosterone as noted) and is the molecular ion (M) for many compounds. The transition product quantitation (Quant) ion and the primary (Q1) and secondary (Q2) qualifying ions are shown along with the argon gas collision energy (CE) used for each transition. Additional used or unused (in *italics*) qualifying ions (Q3 and Q4) and their optimized CE, if determined, also are shown. Compounds are listed in ascending gas chromatography retention time as shown in figure 6. eV, electron volts; IDS, isotope-dilution standard; IIS, injection internal standard; min, minutes; nd, not determined; RT, retention time; --, not monitored; ion values in atomic mass units]

Analyte, IDS, surrogate or IIS	RT (min)	M+	Precursor ion	Quant ion	CE (eV)	Q1 ion	CE1 (eV)	Q2 ion	CE2 (eV)	Q3 ion	CE3 (eV)	Q4 ion	CE4 (eV)
Bisphenol A-d_{16} [a]	10.62	386.3	368.2	**197.1**	20	296.1	20	--	--	--	--	--	--
Bisphenol A	10.75	372.2	357.2	**191.1**	18	175.1	15	251.1	15	--	--	--	--
Diethylstilbestrol-d_8	12.57	420.3	420.3	**374.2**	22	220.1	18	--	--	*389*	*nd*	--	--
trans-Diethylstilbestrol	12.65	412.2	412.2	**217.1**	18	179.1	20	231.2	18	*383.2*	*nd*	*368.2*	*22*
cis-Androsterone-d_5	17.39	439.3	439.3	**334.2**	12	244.1	19	--	--	--	--	--	--
cis-Androsterone	17.46	434.3	434.3	**329.3**	14	239.2	18	169.2	20	--	--	--	--
Chrysene-d_{12} (IIS)	18.12	240.2	240.2	**240.2**	2	236.2	2	--	--	--	--	--	--
Epitestosterone	20.61	432.2	432.2	**301.2**	17	327.2	19	209.2	19	--	--	--	--
17-*alpha*-Estradiol	21.16	416.3	416.3	**285.2**	18	326.2	6	232.2	15	*244*	*nd*	*129*	*nd*
Nandrolone-d_3	21.35	421.3	421.3	**194.3**	15	182.3	14	--	--	--	--	--	--
Dihydrotestosterone	21.49	434.3	434.3	**195.2**	16	143.2	16	187.2	15	*405*	*nd*	--	--
4-Andostene-3,17-dione[b]	21.97	430.3	430.3	**260.2**	14	169.2	20	209.2	14	245.2	14	181.1	14
Estrone-$^{13}C_6$	22.11	420.3	420.3	**314.3**	17	404.3	17	--	--	--	--	--	--
Estrone	22.11	414.2	414.2	**155.2**	17	309.3	17	231.2	21	--	--	--	--
Testosterone	22.64	432.3	432.3	**209.2**	14	301.2	18	327.2	14	--	--	--	--
Equilin	22.93	412.2	412.2	**182.1**	23	231.2	23	307.2	16	*168*	*nd*	--	--
17-*beta*-Estradiol-$^{13}C_6$	22.99	422.3	422.3	**288.3**	15	332.3	15						
17-*beta*-Estradiol	22.99	416.3	416.3	**285.3**	16	232.2	15	129.1	15	*326.2*	*6*	*244*	*nd*
11-Ketotestosterone	24.00	518.3[c]	503.3[d]	**323.2**	12	169.1	15	503.3	10	359.2	15	*372.2*	*16*
Norethindrone	25.31	442.3	442.3	**302.3**	10	287.2	17	194.2	19	*233*	*19*	--	--
Mestranol-d_4	25.38	386.3	371.3	**195.1**	16	169.1	16	--	--	--	--	--	--
Mestranol	25.50	382.2	367.2	**193.1**	16	167.1	16	173.1	16	--	--	--	--
Equilenin	26.31	410.2	395.2	**305.2**	8	279.2	12	213.2	17	*168*	*nd*	*181*	*nd*
Ethynylestradiol-d_4	26.78	444.3	429.3	**195.2**	19	233.2	19	--	--	*198*	*nd*	*287*	*nd*
17-*alpha*-Ethynylestradiol	26.90	440.3	425.2	**193.2**	19	231.2	19	205.2	17	--	--	--	--
Cholestane-d_6 (IIS)	29.00	378.4	378.4	**121.1**	20	223.2	5	--	--	--	--	--	--
Estriol-d_4	29.32	508.3	508.3	**314.2**	11	300.2	16	--	--	--	--	--	--
Estriol	29.44	504.3	504.3	**311.3**	15	296.2	15	270.2	18	*324*	*nd*	*386*	*nd*
Progesterone-$^{13}C_3$	30.50	461.3	461.3	**447.2**	8	357.2	4	446.2	8	--	--	--	--
Progesterone	30.50	458.3	458.3	**157.2**	19	353.3	12	235.2	13	*299.2*	*13*	*209.2*	*17*
Epiestriol-d_2	30.79	506.3	506.3	**388.3**	8	326.3	10	--	--	--	--	--	--
Medroxyprogesterone-d_3	33.25	563.3	563.3	**318.3**	16	331.3	14	--	--	--	--	--	--
3-*beta*-Coprostanol	34.75	460.4	370.4	**215.2**	8	257.2	10	313.2	8	--	--	--	--
Cholesterol-d_7	36.12	465.4	375.4	**346.2**	9	255.2	9	--	--	*159.1*	*10*	*145.1*	*10*
Cholesterol	36.25	458.4	368.4	**339.2**	9	255.2	9	159.1	10	*145.1*	*10*	--	--

[a]Derivative M is d_{14}.

[b]For 4-andostene-3,17-dione, the more responsive 430.3-to-234.2 precursor-to-product ion transition initially was selected as the quantitation ion, but was subsequently omitted because of interferences observed in some matrices.

[c]Precursor ion used for transitions to Q2, Q3, and Q4 product ions.

[d]Precursor ion used for transitions to Quant and Q1 product ions.

10.4. Mass Spectrometer Tuning

The mass-axis calibration for the Quattro-micro-GC™ (Waters Corp., section 5.17.2) for MS/MS analysis has been found to be stable for many months, and is verified as needed based on ion-tune observations and at least once every 6 months. Instrument operational tune typically remains fairly stable and only minor adjustments are needed on a regular basis. Nevertheless, before analyzing a batch of samples, an instrument tune is verified and the signal is optimized in full-scan mode. Before tuning, sufficient vacuum must be present (0.67 to 1.1 millipascal with argon collision gas turned off). In addition, the air and water spectra are checked for signs of atmospheric contamination (leaks) as indicated by the abundance of the nitrogen and oxygen ions (mass-to-charge ratio [m/z] = 28 and 32, respectively) (1) having a nitrogen-to-oxygen ion abundance ratio similar to the atmospheric ratio of about 4 to 1, or (2) being 20 times greater than the abundance of the water ion (m/z = 18). The instrument is tuned using perfluorotributylamine, and mass-spectrometry parameters, such as repeller and lens voltages, can be adjusted to maximize the signal of characteristic fragment ions of perfluorotributylamine, including m/z = 131, 219, 414, and 502. If this procedure is not successful in achieving satisfactory performance, corrective action (instrument maintenance) is taken before final preparation of any samples and standards waiting to be analyzed to ensure that instrumental analysis can be carried out promptly after derivatization.

10.5. Instrument Calibration

A multiple-concentration calibration is carried out for all analytes in the method. Before injecting the first calibration standard, at least three injections of sample extracts are made to condition the injection system with sample matrix. It is important that all calibration standards contain the same quantity (mass) of each IDS that was added to samples before extraction. The concentration of each calibration standard and the volume of the low or high analyte mixture, IDS, and IIS solutions required to make each standard are given in table 4. At each calibration level, the concentration is 10 times higher for bisphenol A and 100 times higher for cholesterol and 3β-coprostanol, because these compounds typically are present in field samples at concentrations much greater than the other method analytes.

The IDS compounds are calibrated relative to IIS compounds chrysene-d_{12} or cholestane-d_6. Method analytes are calibrated relative to an IDS that is an exact isotopic analog or is structurally similar (table 7). Section 11, "Quantitation and Calculation of Results," provides additional details.

10.6. Sample Analysis Sequence

After the instrument tune and initial calibration are determined to be satisfactory, a sequence of environmental samples, laboratory QC samples (set blank or set spike), and instrument QC samples are analyzed as a batch. A batch typically consists of sets of 10 environmental samples plus one laboratory reagent-water spike and one laboratory reagent-water blank, separated by instrument QC samples, such as continuing calibration verification (CCV) standards (see table 8). Vials for additional QC instrument standards or blanks (performance evaluation and instrument solvent blanks as described in section 13.1.6) can be inserted if available or needed. Due to instability at room temperature of derivatized progesterone, it is important that the total batch run time does not exceed 72 h. This creates a practical limit of four 10-sample sets that can be combined into one batch, with two sets being typical. Regardless of set affiliation, samples that appear cleaner (less extract color) are positioned earlier in the batch analysis, whereas samples from dirty sites (WWTP effluent or influent samples or those with more extract color) are positioned near the end of the batch analysis to minimize potential instrumental carryover. It is important that all samples are bracketed by acceptable instrument QC standards (CCV and instrument detection level standards; see section 13), and, when this is not the case, initial calibration will be repeated and non-bracketed environmental samples will be reanalyzed. Section 13 (QA/QC) provides a description of the types of QC standard and sample types that are used and their performance criteria.

10.7. Use of Isotope-Dilution Standards

This method uses isotope-dilution quantification to enhance the accuracy of determined analyte concentrations by the addition of IDS compounds to all samples and the calibration standards. IDS compounds are direct or structurally similar stable-isotope labeled analogs of method analytes that are added to samples just before extraction and used to monitor method performance and to correct analyte data for any biases derived from poor recovery during sample preparation, incomplete derivatization yield, or signal suppression during instrumental analysis. Typically, IDS compounds differ from method analytes only in that deuterium (D, or "d" in IDS names in this report) or ^{13}C labels are substituted for hydrogen or carbon-12 atoms at various points on the molecule (fig. 2). As a result, chemical properties are nearly identical to the method analytes, especially when using an exact isotopic analog of an analyte, and IDS recovery can be used as a proxy for absolute analyte recovery by the method. The mass difference resulting from label substitution allows the IDS compounds to be discriminated from the analyte when using mass spectrometry.

Functionally, the IDS compounds are similar to surrogate compounds used in other methods implemented at the NWQL and elsewhere. Like surrogates, IDS compounds are added

Table 7. Method analyte and the corresponding isotope-dilution standard (IDS) used for its quantification, and the injection internal standard used to determine the corresponding IDS's absolute recovery.

[Table is sorted based on IDS grouping]

Method analyte	Isotope-dilution standard	Injection internal standard
17-*alpha*-Ethynylestradiol	17-*alpha*-Ethynylestradiol-d_4	Chrysene-d_{12}
17-*alpha*-Estradiol	17-*beta*-Estradiol-$^{13}C_6$	Chrysene-d_{12}
17-*beta*-Estradiol	17-*beta*-Estradiol-$^{13}C_6$	Chrysene-d_{12}
Equilenin	17-*beta*-Estradiol-$^{13}C_6$	Chrysene-d_{12}
Bisphenol A	Bisphenol A-d_{16}	Chrysene-d_{12}
3-*beta*-Coprostanol	Cholesterol-d_7	Cholestane-d_6
Cholesterol	Cholesterol-d_7	Cholestane-d_6
cis-Androsterone	*cis*-Androsterone-d_5[a]; Nandrolone-d_3	Chrysene-d_{12}
trans-Diethylstilbestrol	Diethylstilbestrol-d_8	Chrysene-d_{12}
Estriol	Estriol-d_4[b]; 16-Epiestriol-d_2	Cholestane-d_6
Equilin	Estrone-$^{13}C_6$	Chrysene-d_{12}
Estrone	Estrone-$^{13}C_6$	Chrysene-d_{12}
Progesterone	Progesterone-$^{13}C_3$[c], Medroxyprogesterone-d_3	Chrysene-d_{12}
Mestranol	Mestranol-d_4	Chrysene-d_{12}
11-Ketotestosterone	Nandrolone-d_3	Chrysene-d_{12}
4-Androstene-3,17-dione	Nandrolone-d_3	Chrysene-d_{12}
Dihydrotestosterone	Nandrolone-d_3	Chrysene-d_{12}
Epitestosterone	Nandrolone-d_3	Chrysene-d_{12}
Norethindrone	Nandrolone-d_3	Chrysene-d_{12}
Testosterone	Nandrolone-d_3	Chrysene-d_{12}

[a]*cis*-Androsterone-d_5 was implemented as the IDS for quantifying *cis*-androsterone on October 1, 2011. For the validation data summarized in this report, the non-exact IDS analog nandrolone-d_3 was used as the IDS for quantifying *cis*-androsterone because *cis*-androsterone-d_5 previously was not available.

[b]Estriol-d_4 was implemented as the IDS for quantifying estriol on March 17, 2011. For the validation data summarized in this report, the stereoisomer 16-epiestriol-d_2 was used as the non-exact IDS analog for quantifying estriol because estriol-d_4 previously was not available.

[c]Progesterone-$^{13}C_3$ was implemented as the IDS for quantifying progesterone on March 1, 2012. For the validation data summarized in this report, the non-exact IDS analog medroxyprogesterone-d_3 was used as the IDS for quantifying progesterone because progesterone-$^{13}C_3$ previously was not available.

Table 8. Typical gas chromatography-tandem mass spectrometry analysis sequence.

[--, not applicable]

Injection order	Type	Prep set number for sample
1–9	Calibration standards (0CAL to 1000CAL)[a]	--
10	Instrument solvent blank (ISB; solvent only)	--
11	Performance evaluation blank (PEB; solvent only)	--
12	Third-party check (TPC)	--
13	Laboratory reagent water blank (LRB, set blank, PBLNK)	1
14	LRB	2
15	Laboratory reagent-water spike (LRS, set spike, PSPK)	1
16	LRS	2
17–26	Set of 10 or fewer environmental samples	1
27	Continuing calibration verification (CCV) standard 50CCV (or 100CCV if sample reanalysis required)	--
28–37	Set of 10 or fewer environmental samples	2
38	50CCV (or 100CCV if sample reanalysis required)	--
39	Instrument detection level standard (5IDL)	--
40+	Additional sets of samples as long as CCV and IDL bracket last set	--

[a]Includes a 5CAL calibration standard that is used as the initial instrument detection level standard (5IDL).

before any sample processing and are used to assess method performance on a sample-to-sample basis. However, surrogate recoveries are used simply as a QC-evaluation tool; surrogate performance is not used in other NWQL methods (as of March 2012) to correct analyte concentrations reported by the NWQL. The IDS compounds in this method are used for analyte quantification using the isotope-dilution technique, where each reported analyte concentration is automatically corrected for IDS recovery. Instead of calibrating the GC/MS/MS based on a ratio of the analyte-to-IIS response as is typical for other NWQL organic methods, analyte calibration is based on the ratio of analyte-to-associated IDS response. Analyte concentrations in the sample are then determined by using the ratio of analyte-to-IDS response in the sample extract (see section 11).

Table 7 lists all the compounds determined by this method and which IDS is used for relative quantification for each method analyte. For 10 analytes (7 analytes for the validation data provided in this report as described in this section), an exact d- or ^{13}C-labeled isotopic analog is used for calibration and quantification. For the remaining analytes, an exact isotopic analog was either unavailable or unusable because of D-label instability, insufficient number of D atoms, standard purity, or prohibitive cost. For these remaining analytes, chemically similar IDS analogs are used. The use of a non-exact isotopic analog for IDQ of an analyte can introduce some bias (either positive or negative) in the determined concentration of the corresponding analyte compared with use of an exact isotopic analog, because the absolute recovery of the IDS through the procedural steps might not exactly match the absolute recovery of the determined analyte. However, based on performance results provided in this report, analyte quantification typically was improved by applying the isotope-dilution procedure in comparison to the traditional approach of quantifying analytes using an IIS procedure. Analyte method recoveries obtained by using IDQ were consistently closer to the desired 100-percent optimum in comparison to the lower absolute recoveries observed for the IDS compounds in the validation matrices (see the "Validation Results" section).

The IDS analog 17β-estradiol-$^{13}C_6$ is used as the exact analog to quantify 17β-estradiol, and as the non-exact IDS analog to quantify its stereoisomer 17α-estradiol. Similarly, estriol was quantified using the non-exact stereoisomer IDS 16-epiestriol-d_2 for the performance data presented in this report. On March 17, 2011, 16-epiestriol-d_2 was replaced as estriol's IDS by the exact deuterium analog estriol-2,4,16,17-d_4 (estriol-d_4) to further improve accuracy of quantitation for estriol. Improved performance is based in part on the previous use of estriol-d_3 as estriol's IDS (as described later in this section). Presently (March 2012), 16-epiestriol-d_2 has been retained as a surrogate compound.

17β-Estradiol-$^{13}C_6$ also is used to quantify equilenin, and estrone-$^{13}C_6$ is used to quantify equilin. The labeled androgen nandrolone-16,16,17-d_3 has one ketone group in the 3 position (fig. 2), is structurally most similar to epitestosterone and

testosterone (fig. 1), and is used to quantify all five natural androgens and the synthetic progestin, norethindrone (table 7). [Note: *cis*-androsterone-2,2,3,4,4-d_5 (*cis*-androsterone-d_5), which is the direct isotopic analog of *cis*-androsterone, became commercially available in April 2011 and was implemented as the exact IDS for *cis*-androsterone on October 1, 2011.] Medroxyprogesterone-d_3 was used to quantify progesterone for the performance data presented in this report. It was replaced as the IDS for progesterone on March 1, 2012, by the exact-analog progesterone-2,3,4-$^{13}C_3$. Presently (March 2012), medroxyprogesterone-d_3 has been retained as a surrogate compound. Cholesterol-d_7 is used to quantify 3β-coprostanol.

Several exact-analog IDS compounds were tested during method development and deemed unsuitable for use. Equilin-2,4,16,16-d_4 initially was rejected because its derivative shares the same nominal parent mass (416.3 atomic mass units) as, and co-elutes with, the unlabeled 17β-estradiol derivative. As a result, there was a specific and reproducible interference with 17β-estradiol identification and quantitation when equilin-d_4 was added to samples. Because 17β-estradiol is more biologically active and more prevalent in environmental samples than equilin, it was determined that the marginal benefits associated with enhanced equilin quantitation by use of equilin-d_4 were outweighed by its potential interference with 17β-estradiol determination. Regardless of this interference issue, equilin-2,4,16,16-d_4 ultimately would have been rejected because of deuterium-label instability as discussed below.

The IDS compounds 17α-estradiol-2,4-d_2, *cis*-androsterone-16,16-d_2, and androstenedione-2,3,4-$^{13}C_3$ were rejected because of insufficient purity (and deuterium label instability for *cis*-androsterone-16,16-d_2). Most IDSs were received with isotopic enrichment levels of approximately 98-atom percent. As a result, some quantity of the unlabeled analyte was always present in the labeled analog. This quantity decreases substantially as the number of labeled sites on the compound increases. It was determined that exact d_2 and $^{13}C_2$ IDS analogs of method analytes were not suitable for this analysis and that use of d_3 and $^{13}C_3$ IDS compounds would be marginal. Indeed, estriol-2,4,17-d_3 initially was used in the method, but had to be spiked into samples at 10 percent of the fortification concentration used for the other hormone IDS compounds to minimize unlabeled estriol signal. To eliminate risk of false estriol signal from estriol-d_3 use, the stereoisomer 16-epiestriol-d_2 was substituted as the IDS for quantifying estriol for the validation results presented in this report. More recently, estriol-2,4,16,17-d_4 became commercially available, which allowed for its substitution as the IDS for estriol on March 17, 2011, as described previously in this section.

The IDS 17β-estradiol-2,4,16,16-d_4 initially was tested and used; however, although substantially more expensive, 17β-estradiol-$^{13}C_6$ was substituted for 17β-estradiol-2,4,16,16-d_4 because of improved purity. All exact-analog IDS compounds used in this method for which data are provided in this report had four or more labeled sites and were found to have acceptable purity. Use of 16-epiestriol-d_2, medroxyprogesterone-d_3, and nandrolone-d_3 containing less than four label positions was possible because the corresponding unlabeled compounds are not determined as method analytes.

Six exact-analog compounds initially tested and used in the method (4-androstene-3,17-dione-2,2,4,6,6,16,16-d_7; dihydrotestosterone-1,2,4,5a-d_4; estrone-2,4,16,16-d_4; norethindrone-2,2,4,6,6,10-d_6; testosterone-2,2,4,6,6-d_5, and progesterone-2,2,4,6,6,17a,21,21,21-d_9) were found to be susceptible to D-loss due to deuterium-hydrogen exchange. This exchange occurred in methanol extracts of environmental samples at one or more labeled positions on alpha-carbons adjacent to ketone functionalities through keto-enol tautomerization. The amount of loss increased dramatically if the extracts were heated above ambient temperature during evaporation steps, but also was found to occur at slower rates even in the IDS solutions in methanol stored primarily at −15°C (Foreman and others, 2010). Therefore, these six isotopes were removed from the method beginning with samples prepared in 2010. The validation data summarized in this report do not include use of these six IDS compounds, except for the analyte holding-time studies, which were conducted just before their elimination. It is noted that deuterium-hydrogen exchange might occur for these six isotopes, or structurally similar isotopes, used in other IDQ methods for hormones that use protic solvents; for example, EPA method 1698 uses norethindrone-d_6 and progesterone-d_9 (U.S. Environmental Protection Agency, 2007a). Deuterium-hydrogen exchange was not observed for the remaining deuterium-labeled IDS compounds in table 7 because their D-labels are not situated adjacent to a ketone functional group.

10.8. Qualitative Determination

The GC/MS/MS system is operated in MRM mode. For each target analyte, at least three precursor-to-product transitions are monitored (table 6). One MRM ion (quantitation ion) is used for quantitative determination of concentration. The other two MRM ions (qualifying ions) are used for unique identification and confirmation of the target analytes. For the IDS compounds, two precursor-to-product transitions are monitored (one quantitation and one qualifying). Compound identification is based on matching the retention time and the ratio of the peak area of the quantitation ion to the peak area for each qualifying ion to the observed values from at least one (mid-concentration level) calibration standard analyzed during the same instrument run. A peak may be identified as present when the following criteria are met.

10.8.1. *Retention time*: The quantitation and qualifying ion peaks elute within ±0.02 min of each other and within ±0.05 min of the retention time observed in a standard. However, some complex sample matrices have substantial amounts of coextracted, chromatographable components that produce increases in the chromatographic retention times of the method compounds. If the IDS associated with a particular analyte has shifted by the same amount as the analyte, the retention time

criterion has been met. Throughout method development and custom analysis, no case was observed where matrix effects caused a compound to elute substantially earlier in a sample than in the calibration standards analyzed during the same run.

10.8.2. *Signal-to-noise ratio*: The minimum signal-to-noise ratio is at least 5-to-1 for the quantitation ion and at least 3-to-1 for the qualifying ions based on either peak-to-peak or root-mean square based values determined using the chromatographic software.

10.8.3. *Ion ratios*: Tandem mass spectrometry is used for this analysis in part for the selectivity of MRM transitions. In general, an unknown peak will have the same major fragmentation pattern as observed in at least one (mid-concentration level) calibration standard. For each unknown peak, the quantitation ion-to-qualifying ion area ratios are measured for both qualifying ions. Three transitions are monitored so that compound identification still is possible even if there is an interference affecting one of the qualifier ions. For positive identification, one of the two ratios will match within specified tolerances shown in table 9 (Antignac and others, 2003).

10.8.4. *Peaks not meeting qualification criteria*: When an unknown peak can not be positively identified as described in sections 10.8.1–10.8.3, its concentration generally is reported as less than the interim reporting level (< IRL) or less than the minimum reporting level (< MRL). However, if the determined concentration is greater than the reporting level (for example, because of an interference), then it may be necessary to raise the analyte's reporting level for the sample to that concentration.

11. Quantitation and Calculation of Results

Analyte calibrations are performed by the TargetLynx™ software (Waters Corp.; section 5.17.4) through use of regression equations. When an analyte has been identified in a sample, the concentration of that analyte will be based on the integrated relative area abundance from the primary quantitation ion of that analyte and the area of the corresponding isotope-dilution standard, and the regression line fitted to the initial calibration using response factors relative to the IDS. The concentration of the IDS in the sample is determined similarly and relative to the quantitation ion for the injection internal standard. However, in this case, an

Table 9. Ion ratio (IR) tolerances for positive identification of unknown compounds (adapted from Antignac and others, 2003).

[≤, less than or equal to; ±, plus or minus; >, greater than; <, less than]

Expected ion ratio	Tolerance (percent)
0.5 ≤ IR ≤ 2	±20
0.2 ≤ IR < 0.5 or 2 < IR ≤ 5	±25
0.1 ≤ IR < 0.2 or 5 < IR ≤ 10	±30
IR < 0.1 or IR > 10	±50

average response factor procedure is used because the IDS concentration in each calibration standard does not change (table 4). Subsequent data processing is accomplished by using a software program (MaDCU) written at the NWQL.

11.1. Sample Volume

The volume of water extracted (S_e), in milliliters, is calculated using the following equation:

$$S_e = (Vi - Vf) \tag{1}$$

where

Vi = initial weight of sample (S1) and sample bottle (B), in grams (is identical to ($\equiv$) milliliters) (section 9.6.2); and

Vf = final weight of unextracted sample (if any, Su) and sample bottle (B), in grams ($\equiv$ milliliters) (Note 3 in section 9.7.7 and section 9.7.10).

Note: This procedure assumes that the volumetric density of a typical freshwater sample is 1 gram per milliliter, and, thus, the mass in grams of the sample is assumed identical to the sample volume in milliliters. For samples collected from saline environments, a salinity or density determination can be made and a volume correction applied.

11.2. Isotope-Dilution Standard Quantitation

Each derivatized standard in the 8-level calibration curve (table 4; 1CAL–1000CAL) is amended with the same mass of IIS and, most importantly, the same mass of IDS compounds (from the same lot number of IDS mixture solution) as was added to all of the samples (see Note 1 in section 7.1.4 and section 7.1.8). The IDS compounds are quantified relative to the IIS compound shown in table 7 by first calculating an average of the IDS-to-IIS response factors determined in each calibration standard:

$$RF_{IDS} = (A_{IDS}/A_{IIS})/(C_{IDS}/C_{IIS}) \tag{2}$$

where

RF_{IDS} = response factor for the IDS compound in each calibration standard;

A_{IDS} = integrated peak area of IDS quantitation ion in the calibration standard;

A_{IIS} = integrated peak area of IIS quantitation ion in the calibration standard;

C_{IDS} = concentration of IDS in the calibration standard, in picograms per microliter; and

C_{IIS} = concentration of IIS in the calibration standard, in picograms per microliter.

All calibration standards are used to calculate the average response factor for each IDS.

Concentrations of IDS compounds in sample extracts are calculated relative to the response for the IIS in the sample extract by using the following equation:

$$E_{IDS} = \{A_{IDS}(C_{IIS}/A_{IIS})\}/RF_{IDSavg} \tag{3}$$

where

E_{IDS} = concentration of IDS compound in sample extract, in picograms per microliter;

A_{IDS} = integrated peak area of IDS quantitation ion for the sample extract;

A_{IIS} = integrated peak area of IIS quantitation ion for the sample extract;

C_{IIS} = concentration of IIS in the sample extract (which is the IIS concentration in the N-methyl-N-trimethylsilyl trifluoroacetamide (MSTFA)/IIS mixture; see section 7.1.8), in picograms per microliter; and

RF_{IDSavg} = average of the IDS response factors calculated for each calibration standard by using equation 2.

11.3. Isotope-Dilution Standard Recovery

The absolute method recovery (in percent) for each IDS compound is reported along with analyte concentration data to NWIS, and is calculated in each sample by using the following equation:

$$R_{IDS} = 100 \times \frac{E_{IDS} \times V_{extract}}{K_{IDS} \times V_{IDS}} \tag{4}$$

where

R_{IDS} = recovery of IDS compound in samples, in percent;

E_{IDS} = concentration of IDS compound in sample extract, in picograms per microliter;

$V_{extract}$ = final volume of extract, in microliters; Note: typically 200 µL (see section 9.15);

K_{IDS} = IDS compound concentration in IDS mixture (see section 7.1.4), in picograms per microliter; and

V_{IDS} = volume of IDS mixture added to sample (see section 9.6), in microliters.

11.4. Isotope-Dilution Standard Recovery for Samples with Unextracted Portion

Rarely, high concentrations of colloidal particles in water samples can clog the solid-phase extract disk; this leads to partial extraction of a sample that has been fortified with the IDS compounds (see Note 3 in section 9.7.7). In this case, only a portion of the total sample will be extracted and the following IDS recovery calculation is used to account for the unextracted portion:

$$RU_{IDS} = R_{IDS} \times (S_e + S_u)/S_e \qquad (5)$$

where

RU_{IDS} = corrected recovery of IDS compound in sample that accounts for an unextracted portion of sample, in percent;

R_{IDS} = uncorrected recovery of IDS compound in sample from equation 4, in percent;

S_e = sample volume extracted, in milliliters; and

S_u = sample volume not extracted, in milliliters.

Note: Correction for an unextracted sample volume (S_u) is relevant for the IDS recovery only and is not required for determining method analyte concentrations. For a well-mixed IDS-fortified sample, the ratio of analyte-to-IDS does not change regardless of the amount of sample extracted.

11.5. Analyte Calibration Curves

For analyte calibration, the relative quantitation ion area response of the analyte to its IDS compared to the relative concentration is calculated by the following equation (for a linear regression model):

$$A_z/A_{IDS} = a(C_z/C_{IDS}) + b \qquad (6)$$

where

A_z = integrated peak area of analyte (z) in each calibration standard;

A_{IDS} = integrated peak area of IDS quantitation ion in each calibration standard;

C_z = concentration of analyte (z) in each calibration standard, in picograms per microliter;

C_{IDS} = concentration of IDS in each calibration standard, in picograms per microliter;

a = slope of linear regression model; and

b = y-intercept of linear regression model.

Note: a similar calculation can be made for fitted quadratic-curve calibrations by the equation $A_z/A_{IDS} = a(C_z/C_{IDS})^2 + b(C_z/C_{IDS}) + c$, where a, b, and c are experimental constants determined from the fitted curve by iterative mathematical extraction with curve fitting software.

For most method analytes, a linear regression model that is weighted ($1/X$ or $1/X^2$) towards the low end of the curve, where X is the concentration level of the calibration standard, and that includes the origin as a fit point usually describes well the relative concentration-to-relative-area response for the calibration standards. Cholesterol and 3β-coprostanol are calibrated at concentration levels that are 100 times higher than the other compounds and require quadratic fits with $1/X$ weighting. The y-intercept is to be less than the instrumental detection level for each compound and near zero. The curve-fit measurement R^2 is to be 0.99 or greater. The determined analyte concentration in the standard is not to exceed ±25

percent of the expected concentration, except in the lowest two calibration standards, which is to be within 30 percent of expected, if detected.

To further reduce their influence on the model curve fit, the upper-most calibration levels can be excluded from the calibration if the analyte concentration in all the samples is less than the concentration of the highest calibration concentration that is retained in the model calculation. However, at least five calibration levels are to be used to define the regression model. Calibration points that are not at an extreme end of the curve are not to be dropped without evidence of some sort of documented failure in preparation or analysis. Additional calibration performance details and corrective actions are provided in NWQL SOP ORGM0477.x, "Analysis of hormone samples by GC/MS/MS—Laboratory schedules 2434, 4434, 6434, and 7434" (Chris Lindley and others, written commun., 2011).

11.6. Analyte Quantitation in Extract

Once a target analyte has been qualitatively identified in a sample, the determined concentration of that analyte in the extract (Czx) will be based on the ratio of the integrated peak area from the quantitation ion of that analyte to the area of the appropriate IDS's quantitation ion (table 7). For those analytes using a simple linear-regression calibration model, the extract concentration is determined by rearrangement of equation 6 to give:

$$C_{zx} = C_{IDSx}\{(A_{zx}/A_{IDSx}) - b\}/a \qquad (7)$$

where

C_{zx} = concentration of analyte (z) in the sample extract (x), in picograms per microliter;

A_{zx} = integrated peak area of analyte in the sample extract;

A_{IDSx} = integrated peak area of IDS quantitation ion in the sample extract;

C_{IDSx} = fortified concentration of IDS in the sample extract, in picograms per microliter, calculated by using the following equation (see terms in equation 4):

$$C_{IDSx} = K_{IDS} \times V_{IDS}/V_{extract} \qquad (8)$$

11.7. Analyte Concentration in the Sample

The concentration of analyte in the water sample is calculated by using the following equation:

$$C_{sample} = \frac{m_{sample}}{S_e} = \frac{C_{zx} \times V_{extract}}{S_e} \qquad (9)$$

where

C_{sample} = concentration of analyte in sample, in

picograms per milliliter (= nanograms per liter, the reporting unit);

m_{sample} = mass of analyte in sample, in picograms.

11.8. Procedure and Calculations for Dilutions

Analyte concentrations are to be within the range of the calibration curve, except as described for cholesterol and 3β-coprostanol in this section. The analyte-to-IDS ratio used in quantitation is not altered by simple dilution. Therefore, the procedure for calculating analyte concentration(s) in diluted samples deviates from that applied in other NWQL methods. Samples are diluted as appropriate using the standard injection solvent (MSTFA/IIS). Dilutions are made by adding 240 μL of MSTFA/IIS solution (section 7.1.8) directly to a GC vial using a syringe, and then by adding 10 μL (a 25× dilution factor), 5 μL (49×), or 2.5 μL (97×) of the undiluted sample extract (at 200 μL volume; see section 9.15.3) with a 10-μL syringe depending on amount of dilution required. The target concentration of the diluted sample is 20 percent or less of the highest calibration standard to ensure that the diluted concentration falls within the dynamic range of the calibration. The dilution vial is capped and mixed on a vortex mixer. The dilution is analyzed as part of the primary batch analysis or in a subsequent batch analysis.

Dilutions are best prepared and analyzed before the undiluted extract as a part of the primary batch analysis for those sample matrices, like WWTP influents, anticipated to have undiluted analyte concentrations greater than the calibration curve. The undiluted sample extract is analyzed for complex matrices where extract dilution is anticipated for two reasons: (1) to provide the IDS recovery value for the sample that is subsequently used to calculate the final IDS recovery-corrected concentration for the diluted analyte (this is necessary because dilution typically makes direct isotope-dilution quantification impossible because the IDS instrumental response is too low or non-existent in the diluted extract depending on the amount of dilution), and (2) to allow for determination of the remaining analytes that might be present at low concentration but would not be detectable in the diluted extract.

Cholesterol and 3β-coprostanol might be present beyond the method's calibration range because they commonly occur at concentrations that are orders of magnitude higher than the concentrations of the steroid hormones and other method analytes in some matrices (for example, wastewater). Sample extracts are not required to be diluted solely because concentrations of one or both of these sterols is out of range because this method was designed specifically to determine 17 of the target analytes (especially the steroid hormones) at very low concentrations. In this case, the data for cholesterol or 3β-coprostanol, or both, can be reported as greater than the highest calibration-level concentration. However, if dilution of a sample is warranted because one of the other analytes is present above the method's calibration range, it is appropriate to apply the same dilution-based calculation to cholesterol and

3β-coprostanol and report those data without qualification if the dilution falls within the calibration range.

The intermediate concentration for the analyte requiring dilution is quantified relative to chrysene-d_{12} or cholestane-d_6 (cholesterol, 3β-coprostanol, and estriol only) instead of the IDS normally used (table 7). This intermediate concentration is then divided by the fractional IDS recovery (equation 3) determined in the original analysis of the undiluted sample extract to account for any analyte loss during sample preparation. It is then multiplied by the dilution factor to reach a final extract concentration (in picograms per microliter) that is used to calculate the analyte concentration in the sample using equation 9. The undiluted sample extract is analyzed to determine concentrations of those analytes present at low concentrations even for sample matrices anticipated to require dilutions for other compounds.

12. Reporting of Results

12.1. Reporting Units

Analyte concentrations for field samples are reported in nanograms per liter to no more than two decimal places (one-hundredths place) and generally to no more than three significant figures. Isotope-dilution standard data for each sample type are reported as percent recovered to one decimal place (tenths of a percent), but to no more than three significant figures. Data for the laboratory reagent-water spike sample are reported as percent recovered to one decimal place, but to no more than three significant figures. Analytes quantified in the laboratory reagent-water blank sample are reported in nanograms per liter to two decimal places, but to no more than three significant figures.

12.2. Reporting Levels

Estimated detection and reporting level values and reporting level types applicable on October 1, 2011 are summarized in table 10. (Note: Several of these detection or reporting levels differ from the values applied to the performance data presented in this report, as described in the "Method Validation and Additional Performance Data" section.) Sixteen of the method analytes are reported using the laboratory reporting level (LRL) convention as described in Childress and others (1999) using interim reporting levels (IRLs) set at twice the applied detection levels for most analytes (as described in the "Assessment of Blank Contamination and Determination of Detection and Reporting Levels" section). Because qualitatively identified detections that are less than the detection level can provide useful information (Childress and others, 1999), concentrations for these 16 analytes that are less than the detection level or less than the lowest calibration standard are reported to NWIS with one or more result-level value qualifier codes as described in Office of Water Quality Technical Memorandum 2010.07 (U.S. Geological Survey, 2010). Compounds that are not detected or that do not meet qualitative criteria are reported as less than the reporting level. Matrix-specific interferences might warrant the reporting of raised reporting levels.

Bisphenol A, cholesterol, and 3β-coprostanol are sample-preparation blank-limited analytes, and 11-ketotestosterone is an instrumental blank-limited analyte (see "Blank-Limited Analytes" section). These four analytes are reported to NWIS using the minimum reporting level convention (Childress and others, 1999). The MRL is the smallest measured concentration of a constituent that may be reliably reported by using a given analytical method (Timme, 1995); no data are reported at concentrations less than the MRL.

Table 10. Detection and reporting levels used for reporting analyte concentrations by the analytical method to the U.S. Geological Survey National Water Information System (NWIS).[a]

[NA, not applicable; ng/L, nanograms per liter]

Method analyte	Detection level (ng/L)	Reporting level[b] (ng/L)
11-Ketotestosterone	NA	2
17-*alpha*-Estradiol	0.4	0.8
17-*alpha*-Ethynylestradiol	0.4	0.8
17-*beta*-Estradiol	0.4	0.8
3-*beta*-Coprostanol	NA	200
4-Androstene-3,17-dione	0.4	0.8
Bisphenol A	NA	**100**[c]
Cholesterol	NA	200
cis-Androsterone	0.4	0.8
Dihydrotestosterone	2	4
Epitestosterone	**1**[c]	**2**
Equilenin	1	2
Equilin[d]	**4**	**8**
Estriol	1	2
Estrone	0.4	0.8
Mestranol	0.4	0.8
Norethindrone	0.4	0.8
Progesterone[d]	4	8
Testosterone	**0.8**	**1.6**
trans-Diethylstilbestrol	0.4	0.8

[a]Detection and reporting levels shown are applied to sample data provided using National Water Quality Laboratory schedules 2434 and 4434 as of October 1, 2011.

[b]All analytes are reported using the laboratory reporting level (LRL) convention with information-rich data reporting and a NWIS interim reporting level (IRL) type code, except for 11-ketotestosterone, bisphenol A, cholesterol, and 3-*beta*-coprostanol, which are reported using a minimum reporting level (MRL type code) convention; see Childress and others (1999) and section 12.2.

[c]Detection and reporting level values shown in **bold** differ from the detection and reporting level values that were applied to the performance data presented in this report; see "Assessment of Blank Contamination and Determination of Detection and Reporting Levels" section.

[d]Concentrations for this analyte are reported as estimated only (NWIS "E" remark code for estimated concentration).

12.3. Data Qualification Criteria

Definitions for the NWIS data codes described in this section are documented in U.S. Geological Survey (2011a). All concentration data for equilin and progesterone are reported as estimated-only (NWIS "E" result-level remark code) because they exhibit excessive bias or variability, or both in some matrices (see "Method Validation and Additional Performance Data" section). For the remaining 18 method analytes, the concentration (analytical result) in a given sample is reported based on the sample-specific recovery of the corresponding IDS relative to the performance criteria shown in table 11. The analyte concentration (typically) is reported without an "E" remark code if the IDS is an exact isotopic analog of the analyte and the IDS recovery is in the 25–120 percent range. For analytes that use a non-exact IDS analog, the "E" remark code (typically) is not applied to the analyte result if the IDS recovery is in the 40–120 percent range (table 11). Note that an analyte's result might include the "E" remark coded for reasons other than IDS recovery performance (Childress and others, 1999).

If the IDS recovery is less than 5 percent, the analyte concentration is not reported, regardless of whether the analyte is detected or not. Instead, one of following three result-level null-value NWIS codes is reported to NWIS: (1) the "M" remark code (NWIS description "Presence of material verified but not quantified") is reported if the analyte was detected and the recoveries for the other IDS compounds generally were greater than 5 percent; (2) the "r" qualifier code (NWIS description "Sample ruined in preparation") is reported if the recoveries for many of the IDS compounds in the sample are less than 5 percent (or possibly slightly higher) and the sample is believed to have been ruined; or (3) the "x" code (NWIS desription "Result failed quality-assurance review") is reported if the analyte was not detected and the recovery for its corresponding IDS compound was less than 5 percent, but the recoveries for other IDS compounds in the sample generally were greater than 10 percent.

Both the "r" and "x" codes signify a quality-assurance/quality-control failure. Use of the "x" code as an analyte-specific qualifier is applied for this method because of IDS use. Application of the "x" code is preferred relative to the "r" null-value qualifier code that historically has been used by the NWQL when specific analyte data in a sample are not reported because of performance issues that compromise reporting of quantitative results or reliable application of a reporting level. (Note: the "r" code historically has been applied by the NWQL as the default analyte-specific null-value qualifier code even though data for other analytes in the sample are reported because, in fact, the entire sample is not "ruined.") Additional or alternative NWIS codes to those shown in table 11 might be applied to reported results, as appropriate. Several additional data-reporting considerations are described in section 13.

Table 11.   Criteria used for application of National Water Information System (NWIS) result-level codes to reported analyte data based on sample-specific recovery of the isotope-dilution standard (IDS).[a]

[E, estimated remark code; M, null-value remark code defined as "presence of material verified but not quantified;" r, null-value (result level) qualifier code defined as "sample ruined in preparation"; x, null-value (result level) qualifier code defined as "result failed quality assurance review;" >, greater than; <, less than]

IDS recovery range (percent)	Applied NWIS code for:	
	Analytes that have an exact IDS[b]	Analytes that use a non-exact IDS[c]
>120	E	E
40 to 120	none	none
25 to <40	none	E
5 to <25	E	E
<5[d]	M, r, or x	M, r, or x

[a]NWIS codes are defined in Appendix A of U.S. Geological Survey (2011a). Coding applied based on information in this table might be in addition to one or more NWIS codes applied to the result for other reasons. Alternative NWIS codes might be applied, as needed.

[b]Analytes quantified using an exact IDS as of March 1, 2012, are 17-*alpha*-ethynylestradiol; 17-*beta*-estradiol; bisphenol A; *cis*-androsterone; cholesterol; estriol; estrone; mestranol; progesterone; and *trans*-diethylstilbestrol. All progesterone concentrations are reported as estimated (see section 12 3).

[c]Analytes quantified using a non-exact IDS as of March 1, 2012 are 11-ketotestosterone; 17-*alpha*-estradiol; 3-*beta*-coprostanol; 4-androstene-3,17-dione; dihydrotestosterone; epitestosterone; equilenin; equilin; norethindrone; and testosterone. All equilin concentrations are reported as estimated (see section 12.3).

[d]There are three analyte coding options for this IDS recovery condition: (1) the "M" code is reported if the analyte is detected; (2) the "r" code is applied if the recoveries for many of the IDS compounds in the sample are <10 percent; (3) the "x" code is applied to the analyte if the recovery of the corresponding IDS compound is <5 percent, but recoveries for other IDS compounds in the sample are >10 percent.

13. Quality Assurance/Quality Control

Key aspects of QA/QC are provided in this section. Additional details are provided by Maloney (2005) and NWQL SOP ORGM0477.x, "Analysis of hormone samples by GC/MS/MS—Laboratory schedules 2434, 4434, 6434, and 7434" (Chris Lindley and others, written commun., 2011).

13.1. Quality-Control Types, Performance Criteria, and Corrective Actions for Instrumental Analysis

The following QC types, performance criteria, and corrective actions are applied to the GC/MS/MS analysis:

13.1.1. *Batch sequence QC considerations*: Environmental and QC samples usually are grouped together for analysis in a batch sequence. A typical GC/MS/MS sequence used to analyze two sets of environmental samples might order the samples and standards as shown in table 8. Most commonly, a calibration curve is analyzed at the beginning of each batch because the calibration standards and two (commonly)

preparation sets of samples are derivatized at the same time. Environmental samples are postioned in the sequence such that they are bracketed by calibration-based performance standards (that is, the initial calibrations standards and subsequent continuing calibration verification standards analyzed throughout the sequence; table 8). Results for these standards indicate whether the GC/MS/MS is providing acceptable calibration performance for those samples that fall between the bracketing standards (see section 13.1.3). The number of samples that can be analyzed successfully between the CCVs may vary based on instrumental performance during an analysis sequence. The analysis sequence also includes other GC/MS/MS performance sample types as described in sections 13.1.4–13.1.6.

13.1.2. *Calibration criteria*: see section 11.5.

13.1.3. *Continuing calibration verification standards*: CCVs are analyzed during a sample sequence to ensure that the calibration of the GC/MS/MS system remains within acceptance limits (Maloney, 2005). CCV frequency in a batch will consist (at a minimum) of a CCV at (1) the first vial position (assuming the batch analysis does not include a complete set of calibration standards and not considering "matrix" sample pre-batch inlet conditioning injections as described in section 10.5), (2) following every tenth (or less) environmental sample throughout the analysis, and (3) after the last environmental sample in the sequence. The CCV concentration may be any level within the calibration range, but currently (March 2012) the mid-range 50CCV and 100CCV levels are used.

13.1.3.1. *CCV performance criteria*: Individual compounds in both of the CCV standards immediately bracketing environmental samples in the analytical run sequence are relevant in the consideration of CCV acceptance criteria. CCV standards are quantified by using the calibration curve. Acceptable CCV performance is demonstrated if the determined concentration of the method compound in the CCV is within 25 percent of the expected concentration (except for progesterone, which has wider performance limit criteria based on more variable performance). If CCV criteria are not met for more than one compound, then environmental samples that follow the last satisfactory CCV are reanalyzed after appropriate corrective action (typically following GC/MS/MS maintenance procedures) and instrument recalibration, except if the compound was not detected in the sample and the instrument detection level criteria were met (section 13.1.4). If the sample can not be reanalyzed, results reported for compounds that were outside the CCV limits are coded in NWIS as estimated.

13.1.4. *Instrument detection level performance criteria*: The instrument-detection-level standard is used to determine if instrument detection capability is sufficient for the determination of low compound concentrations. The instrument-detection-level standard has a concentration near the average reporting level of about 2 ng/L for 17 of the 20 method analytes (average excludes the analytes bisphenol A, cholesterol, and 3β-coprostanol that are reported using higher minimum reporting level concentrations), which equates to 5 pg/μL in a concentrated extract. Thus, the 5CAL (5 pg/μL; table 4) calibration standard is used for the instrument-detection-level standard. An instrument-detection-level standard is injected at the beginning (during calibration) and end of an analysis sequence, and indicates whether instrument sensitivity has deteriorated during sample analysis to the point where determination of concentrations near the reporting level has been compromised. If analytes can not be qualitatively identified in the instrument-detection-level standard analyzed at the end of a sequence, performance has likely been compromised, and samples are to be reanalyzed after GC or mass spectrometer maintenance is performed.

13.1.5. *Third-party check standard*: A third-party check standard is analyzed after each new calibration curve is generated. The third-party check standard is used to verify the calibration standard concentrations and the integrity of the curve. Concentrations of analytes in the third-party check standard within 30 percent of expected values are acceptable. If the determined concentrations do not meet this criterion, the analyst will check that the calibration and third-party check solution concentrations are correct by (1) preparing and analyzing of new calibration and third-party check standards for cross-check; (2) performing system maintenance and reanalyzing the calibration standards and the third-party check standard, and, if acceptable, continue reanalysis of all batch samples; (3) preparing or obtaining a new third-party check standard to compare with original third-party check standard; (4) preparing new calibration standards; or (5) taking other corrective actions.

13.1.6. *Performance evaluation blank*: The performance evaluation blank and instrument solvent blank (a "wash" blank) are solvent-only (typically dichloromethane) instrument blanks. Because these two types of instrument blanks do not contain any IDS or injection internal standard compounds, only compound response (peak areas) can be reviewed by the analyst for signs of instrument-related contamination or compound carry over during injection of extracts. The performance evaluation blank is analyzed before the analysis of the first environmental samples to ensure that none of the method analytes are detected (see table 8 for typical analysis sequence). Bisphenol A, 3β-coprostanol, and cholesterol might be detected in the performance evaluation blank because these compounds are ubiquitous blank contaminants. If these analytes are detected in the performance evaluation blank, the analyst will check that the peak area for these three analytes is at least 10 times lower than that comparable to a standard that would be just below the MRL. If the response is greater than this threshold, then instrument maintenance is performed and the samples are reanalyzed. Performance evaluation blank (and instrument solvent blank) vials also can be interspersed within a sequence to monitor for analyte carryover during extract injections.

13.2. Quality-Control Components, Performance Criteria, and Corrective Actions for Specific Sample Types

13.2.1. *Injection internal standard*: If an IIS compound's peak area is not within ±50 percent of the mean IIS area for the analytical set, the possibility of extract evaporation (which increases areas) or other influences are to be considered, and IDS compounds in affected extracts are evaluated to determine if they have acceptable recoveries (section 13.2.2). System maintenance and reanalysis might be warranted as indicated by IIS peak shape and response as described in NWQL SOP ORGM0477.x, "Analysis of hormone samples by GC/MS/MS—Laboratory schedules 2434, 4434, 6434, and 7434" (Chris Lindley and others, written commun., 2011).

13.2.2. *Isotope-dilution standards*: The IDS compounds are added to all samples and all instrument standards (but not the performance evaluation blank or instrument solvent blank) to achieve isotope-dilution quantification of corresponding method analytes (table 7); see Note 1 in section 7.1.4 regarding IDS use. IDS recoveries are absolute method recoveries from sample extraction through analysis, and reflect corresponding *absolute* analyte recovery (see "Reagent Water" in the Primary Validation Matrices section). The IDS recoveries are used to monitor for sample-specific preparation errors, and dictate subsequent analyte reporting (see table 11, section 12.3, and the following information in this section). IDS recoveries also are indicators of analyte-detection likelihood in relation to the typical detection levels. That is, very low IDS recoveries typically mean low analyte absolute recovery, so less analyte mass is available for detection.

Performance of the IDS is evaluated in concert with IIS performance to determine whether samples warrant reanalysis or if instrument maintenance is warranted. Low recovery of one or more IDS compounds in a sample compared to normal IIS performance (within expected IIS area range) and to normal performance for other QC samples analyzed in the sequence (especially the set LRS and LRB samples; see section 13.3) usually indicates that the IDS compound(s), and corresponding analyte(s), experienced excessive losses during sample-preparation steps or from matrix-specific effects, and were not related to instrumental analysis; thus, reanalysis is unnecessary. Unless the sample is ruined during sample preparation (NWIS result-level null-value qualifier code of "r" provided as described in section 12.3), IDS recovery data are reported for a sample regardless of determined value (including zero recovery) to assist the customer in understanding the reported information for the corresponding analyte(s).

13.2.2.1. *IDS performance criteria*: IDS data from laboratory reagent-water spikes and laboratory reagent-water blanks are acquired and statistically evaluated to develop acceptance criteria on an ongoing (typically yearly) basis, as is done for surrogate compounds in other organic methods (Maloney, 2005). Based on validation and custom implementation-sample data acquired through January 2011, recoveries for most IDS compounds were reasonable in most environmental samples (see "Isotope-Dilution Standard Performance" in the "Long-Term Estimates of Method Performance" section). However, acceptable analyte method recovery (not *absolute* recovery) typically is obtained in matrix- and reagent-spike samples even when the IDS recovery is less than 60 percent (see "Method Validation and Additional Performance Data" section). Therefore, analyte concentrations are reported as long as the corresponding IDS recovery is 5 percent or greater. Analyte data reporting depends on the absolute IDS recovery in the sample (see criteria shown in table 11 and described in section 12.3).

If IDS recoveries for a given sample are unacceptable, IDS recovery in the associated samples, LRS, and LRB are evaluated, along with any anomalous observations recorded during sample preparation, to ascertain if there is a broader method-performance problem that affected the entire set or batch of samples. In general, if no obvious indications of process failure can be attributed to sample preparation or analysis problems, an IDS compound's recovery failure may be attributed to matrix problems, and the results for any detected analytes are reported as shown in table 11.

13.3. Quality-Control Samples, Performance Criteria, and Corrective Actions for Sample Preparation Sets

13.3.1. *Laboratory reagent-water blank*: One LRB sample is included with each sample preparation set and processed in parallel with the associated environmental samples. The LRB sample is prepared using about 450-mL of reagent (typically Solution 2000) water (see section 9.6). The LRB is used to monitor for interferences and the possible introduction of method analytes during sample preparation. Concentrations of analytes detected in the LRB are reported in ng/L (see section 13.8).

Three analytes (bisphenol A, cholesterol, and 3β-coprostanol) are ubiquitous blank contaminants, and data for these analytes are censored below the MRL to avoid the reporting of false positives associated with laboratory contamination. If any of the 16 analytes that are not reported using the MRL convention (see section 12.2) are detected in an LRB, then the analyst will evaluate the possibility that a portion or all of the analyte concentration in the associated environmental samples might result from laboratory contamination.

Typically, if an analyte is detected in an LRB, the concentration is lower than the IRL and even the detection level. Blank detections less than the IRL are possible for this analytical method because mass-spectrometric analysis can result in a qualitatively identified detection whose concentration is less than the statistically-derived detection and reporting levels. Childress and others (1999) and U.S. Geological Survey (2010) provide an explanation of the conventions used to report analytical data below the reporting

level. Samples associated with a contaminated LRB are evaluated as to the best corrective action for the affected samples. The concentration of the detected compound in the LRB is used to qualify or censor, if needed, the concentration in environmental samples using the data reporting procedures described in Office of Water Quality technical memorandum 12.01 (U.S. Geological Survey, 2011c). The concentrations of method analytes detected in LRB samples are not subtracted from those in environmental samples by the NWQL.

13.3.2. *Laboratory reagent-water spike*: One LRS sample is included with each sample set and processed in parallel with the associated environmental samples. The LRS is prepared by fortifying the same reagent-water media used for the LRB with 12.5 ng of 17 of the 20 method analytes (25 ng/L, assuming a 500-mL sample volume, as described in section 9.6). Analyte data for the LRS are reported in percent recovery (see section 13.8). The LRS recoveries track method performance in a reagent matrix that does not include the potentially interfering compounds that might be present in field-sample matrices.

LRS analyte method recoveries are automatically corrected for procedural losses by use of the IDQ procedure and, therefore, typically will be greater than (and, thus, not match) the absolute recoveries for the corresponding IDS compounds in the LRS. For example, the *method* recovery of 17β-estadiol in a given LRS might be 97 percent, whereas the corresponding *absolute* recovery of its IDS, 17β-estadiol-$^{13}C_6$, in that LRS might be 68 percent.

LRS analyte recoveries are used in part to determine if overall recoveries within the preparation set or instrumental batch are acceptable, or if a substantial change in method performance occurred for the set. Individual analyte recoveries in the LRS are interpreted in the context of a historical group of LRS recoveries (Maloney, 2005). At a minimum, this group will consist of 30 or more LRS samples, analyzed over a period of 6 months or more, and processed by multiple analysts, if applicable. Statistical-process-control analysis is applied to these data to develop recovery acceptance criteria. If the recoveries of a set-specific LRS are not acceptable (that is, recoveries are not within three standard deviations of the long-term mean recovery; or similar non-parametric-derived criteria), other measures of set-specific performance, such as IDS recoveries in the environmental samples and LRB for that sample set, also are evaluated to determine if there is a common recovery problem.

First, any observations recorded during preparation of the samples in the set are reviewed. If poor recovery in the set LRS resulted from a sample-processing error, then the analyst will determine whether the error also adversely affected the associated environmental samples and will apply corrective actions or data qualifications as appropriate. If IIS response and, especially, IDS recoveries in the environmental samples for that sample set are acceptable, then results for the environmental sample detection are reported unqualified. If some or all IDS recoveries in samples also do not meet criteria, then relevant analyte data in the environmental sample are reported based on the coding criteria in table 11, or are deleted (NWIS result-level null-value qualifier code of "x" reported). Unusually low recoveries for many IDS compounds might result in an entire sample being reported as ruined (NWIS result-level null-value qualifier code of "r" reported).

13.4. Field-Submitted Environmental Quality-Assurance Samples

U.S. Geological Survey personnel submit field-based QA samples to the NWQL as part of their QA project plan, including various types of field blanks and sample replicates that are treated identically to environmental samples; these include the following:

13.4.1. *Field blanks*: Submission of field blanks is especially warranted for this analytical method (see section 4). Different types of field-submitted blanks are designed to assess blank contamination from field processing procedures as described by Wilde and others (2004), of which only the field blank is specifically noted in this report. The field blank consists of a volume of reagent (analyte free) water that is processed in exactly the same manner as environmental samples by using all appropriate on-site sampling equipment and techniques. This process includes bottles, compositing, splitting, and, for samples submitted for LS 2434, field filtering. The field blank is collected and processed according to the QA plan designed for a given field study. An initial field blank typically is collected and processed at the start of sampling, following equipment cleaning, and then additional field blanks are collected for about every 10 to 20 environmental samples, or more frequently. The field blank, when compared relative to the LRB, helps monitor for contamination or carryover, or both, resulting from field sampling and equipment-cleaning techniques that could cause equipment contamination of environmental samples. Field equipment cleaning procedures for organic contaminants as described in Wilde and others (2004) are suitable for the analytes determined by this method.

13.4.2. *Field replicates*: Field-replicate samples are used to assess within-matrix variability in analyte concentrations (Wilde and others, 2004). Replicates—environmental samples collected in duplicate or higher multiples—are considered identical in composition and are best prepared by collecting the entire required volume of water as a composite sample, and then splitting the sample. For LS 2434 samples, the splitting can be completed during or following the filtration step. Ideally, the relative percent difference (RPD; see equation 10) for duplicate analyte concentrations will be ≤30 percent (note: this criterion matches that used in USEPA method 539 (U.S. Environmental Protection Agency, 2010a)), unless the concentrations are near the detection level when an analyte might not be detected in one of the replicates or exhibit greater variability. RPDs greater than 30 percent might indicate greater matrix-specific variation.

$$RPD = 100 \times \frac{|C_1 - C_2|}{(C_1 + C_2)/2} \qquad (10)$$

where

$C =$ concentration of analyte in duplicate samples *1* and *2*, in nanograms per liter.

13.4.3. *Field-requested laboratory matrix-spike sample*: A FRLMS is obtained by submitting a duplicate environmental sample with a request for laboratory code 4000 (laboratory matrix spike) on the NWQL's Analytical Services Request form. The duplicate is spiked at the NWQL with the method analytes using the same or greater volume of the spike fortification mixture used for the LRS (see section 9.6). Greater spike amounts are warranted for matrices with anticipated higher unspiked concentrations of the analytes, but this must be noted under the "Comment to NWQL" section on the Analytical Services Request form. The FRLMS is prepared and analyzed along with the unfortified sample. Determined analyte concentrations (not percent recoveries) in the FRLMS are reported to NWIS. The USGS data user uses the concentration data from NWIS for the FRLMS and corresponding unspiked sample, along with spike solution lot compositional information provided by the NWQL and the volume of spike solution fortified, to calculate analyte recoveries in the FRLMS (see NWQL Technical Memorandum 2005.02 at *http://nwql.usgs.gov/ Public/tech_memos/nwql.2005-02.pdf*). Recoveries of method analytes in an FRLMS (and MSPK as described in section 13.5.2) can be compared to other reagent-water and matrix-spike performance data provided in this report (see relevant data tables in "Method Validation and Additional Performance Data" section).

13.4.4. *Field matrix-spike sample*: Currently (March 2012), a spiking solution for the preparation of field-matrix-spike samples is unavailable for this analytical method. Similar to the FRLMS, determined analyte concentrations in a field matrix-spike sample are reported to NWIS, and the customer calculates analyte recoveries. Procedures for field spiking are provided in Wilde and others (2004).

13.5. Other Environmental Quality-Assurance Samples

13.5.1. *Laboratory matrix duplicate sample*: The DUP is a replicate field sample randomly selected by NWQL sample-preparation staff for use as a laboratory duplicate. Inclusion of a laboratory duplicate by the NWQL is optional and is possible if additional backup replicate samples are submitted by field staff (at their discretion) or by request from the NWQL for use in method-performance testing. The DUP sample is prepared and analyzed along with other field samples in the sample preparation set. The DUP is assigned the associated environmental sample's NWQL identification number appended with "DUP." DUP data are not reported to NWIS, but are available from the NWQL for use by both the NWQL and field personnel to evaluate matrix-specific variation in analyte concentrations for environmental sample replicates. As noted previously in section 13.4.2 for field replicates, the RPD for detected compounds in the DUP and its corresponding ambient replicate ideally will be ≤30 percent, unless the concentrations are near the detection level when the analyte might not be detected in one of the replicates or exhibit greater variability.

13.5.2. *Laboratory matrix-spike sample*: The MSPK is a backup, replicate field sample randomly selected by NWQL sample preparation staff to be used as a matrix spike sample. As with the DUP, inclusion of a MSPK by the NWQL is optional and is possible if additional backup replicate samples are submitted by field staff. The MSPK is fortified with the method analytes (typically, but not necessarily, at the same level as used for the LRS) and prepared and analyzed along with other samples in the set. The MSPK sample is assigned the associated environmental sample's NWQL identification number appended with "MSPK." Analyte concentrations are determined in the MSPK as with other environmental samples. Analyte recovery in the MSPK is then calculated by the NWQL by subtracting analyte concentrations in the corresponding unspiked field sample from the MSPK sample concentration. Ideally, MSPK analyte recoveries will be similar to those determined for the LRS (see "Analyte Method Recoveries in Laboratory Reagent-Water Spike Samples" section) and other matrix-spike samples (see "Compound Recoveries in Other Spiked-Matrix Samples" section), but can be biased low or high, particularly if the ambient analyte concentration is near or greater than the fortification concentration. This situation might even produce negative analyte recoveries (which are not reported; instead the "x" delete code is reported) or unusually high recoveries. Matrix interferences or effects also can produce biased results. MSPK data in percent recoveries are not reported to NWIS, but are available from the NWQL.

Note: Use of MSPK and DUP samples is more commonly applied to NWQL sediment organic methods because sufficient sediment material typically is available for use as an MSPK or DUP aliquot, whereas an MSPK or DUP for a water method requires submission of duplicate sample bottles.

13.5.3. *Reference standard samples*: Currently (March 2012), certified, standard, or other reference materials are not available for this analytical method.

13.6. External Quality-Assurance Functions Used to Assess This Method

The USGS Branch of Quality Systems' Organic Blind Sample Project routinely submits spiked reagent-water samples at varying concentrations as an external monitor of method performance. The spike sample is typically submitted as an LS 4434 sample, data for which are directly relevant to LS 2434 performance as well, and the sample is prepared

and analyzed within a normal sample set. The Organic Blind Sample Project provides analyte-specific performance results for these samples along with performance summary reports at *http://bqs.usgs.gov/OBSP/index.html* (accessed April 2012). These results are not described in this report. Spike-sample recovery results provided by the Organic Blind Sample Project are used by the NWQL with other method QA/QC data to evaluate long-term method performance.

13.7. Secondary Data Review

Secondary data review is a critical component of QA of all reported environmental data. An independent chemist, who is qualified to perform this analysis, reviews all results and documentation to verify that the original analyst correctly identified and quantified the method analytes by using available QC data and available documentation on sample preparation and analysis. The analytical results for every environmental sample are subject to secondary data review.

13.8. QA/QC Data Availability

All instrumentation QC sample types (section 13.1) and the LRB and LRS samples (section 13.3) are created as LS 2434 sample types only. However, these QC sample types are relevant to, and associated with, both LS 2434 and LS 4434 field-sample data. Of particular importance to USGS staff are the LRB and LRS samples that are prepared and analyzed with each set of samples. The LRB and LRS samples use the NWIS parameter codes assigned for LS 2434 (tables 1 and 2), although the data for these QC samples reside at the NWQL and are not stored in NWIS. Accordingly, USGS staff need to access long-term summaries (charts, box plots, tables) of LRB and LRS sample data relevant to their LS 2434 or LS 4434 field samples by querying LRB and LRS data using the LS 2434 entry point at the Online NWQL QC Data website (USGS access only). Further information on accessing this data is available by contacting the NWQL at labhelp@usgs.gov.

Method Validation and Additional Performance Data

Data are included in this section for primary validation matrices, long-term estimates of method performance, assessment of blank contamination and determination of detection and reporting levels, holding-time experiments, and compound recoveries in other spiked-matrix samples.

Primary Validation Matrices

Unfiltered replicate samples from the following four validation matrices were used to test method performance: (1) reagent water, (2) surface water collected downstream from a wastewater-treatment plant discharge, (3) secondary WWTP effluent, and (4) primary WWTP effluent receiving no biologicial treatment. Additional information relevant to each matrix is provided in the subsections that follow. The non-reagent-water matrices were selected in part because they were collected from a location affected by municipal wastewater discharge (the surface water) or were part of a WWTP flow path. As such, they provide performance data for difficult sample matrices likely to be submitted for analysis by this method. Unfiltered waters (LS 4434) were used for validation tests because they provide a more challenging test matrix than filtered water (LS 2434) and are a better indicator of overall method performance. Subsamples of each matrix were fortified with method analytes at levels higher than their anticipated ambient (unfortified) concentration for most analytes (except for the primary wastewater effluent matrix) based on prior analysis of samples from similar sites. In addition, four unspiked replicates (referred to as "ambient" replicates) of each of the three non-reagent-water (field) sample matrices were analyzed to determine whether any method analytes were present and at what concentrations.

Approximately 20 L or more of each field-sample matrix were collected. A Teflon® churn splitter was used to subsample approximately 0.45-L aliquots for the surface water and secondary effluent matrices into 0.5-L HDPE bottles. About 40 L of the primary-effluent water matrix was pumped into a 50-L plastic container lined with a Teflon® bag. In an attempt to keep the particle loads uniform between sample bottles, the bottles were filled by pumping the continuously mixed water from the bag using a proportional dispensing procedure. Aliquots of the four ambient primary wastewater effluent replicates were collected throughout the bottle filling process to be representative of the subsampling conditions. Several method analytes (especially cholesterol and 3β-coprostanol) are moderately hydrophobic and partition to suspended particles, so it is necessary to ensure homogeneous splitting of samples for reproducibility.

Eight or nine replicates of each matrix were fortified with 17 of the method analytes to assess analyte recovery at 10 ng/L (also referred to as the "low" level) and 100 ng/L (the "high" level) fortification concentrations assuming a 0.5-L sample volume. In all fortified samples, bisphenol A was added at 10-times higher concentrations and cholesterol and 3β-coprostanol at 100-times higher concentrations than the other analytes because these three analytes tend to occur in the environment at substantially higher concentrations and because they are blank limited (as described in the "Blank-Limited Analytes" section). Fortification at sufficiently high levels relative to the ambient concentration reduces the uncertainty in compound recovery calculations. However, in a number of cases (especially for the primary effluent matrix), the ambient analyte concentrations approached or exceeded the fortification concentration by a substantial amount. The validation samples were fortified with the isotope dilution standard compounds at 100 ng/L, except for cholesterol-d_7, which was added at 10,000 ng/L (see section 10.7). The

exception was the secondary effluent matrix, where bisphenol A-d_{16} was inadvertently spiked at a concentration 100 times higher than intended (see "Secondary Wastewater Effluent" section).

Validation Results

All recoveries are means of the spiked replicates and all concentrations are means of ambient (unspiked) replicates for the given field test matrix unless otherwise noted in this section. Mean analyte percent recoveries and relative standard deviations (RSDs) of recovery at the low- and high-fortification levels in the four test matrices are given in tables 12 and 13, respectively, along with mean absolute recoveries and RSDs of the IDS compounds.

In several matrices, the unfortified (unspiked) samples had detectable "ambient" concentrations of analytes (table 14) that complicate the recovery calculation, which is made by subtracting the mean ambient concentration from the determined concentration in the fortified sample and dividing the resultant concentration by the fortification concentration. For those analytes with a mean ambient concentration that is less than 25 percent of the fortification concentration, the mean spike recoveries are reported unqualified in tables 12 and 13. Due to the presence of relatively high ambient analyte concentrations compared to spike fortification concentrations, potential enhanced bias or variability, or both, are denoted as follows in tables 12 and 13: mean spike recoveries are shown in bold for those analytes with mean ambient concentrations from 25–150 percent of the fortification concentration and in bold italics for those from 151–300 percent of the fortification concentration. In cases where the ambient concentration exceeded 300 percent of the amount spiked, the recovery is either provided to show performance under this condition (denoted by bold italics and footnoted) or no recovery is reported because of extensive bias.

Mean recovery of IDS compounds provides an estimate of absolute analyte recovery because the IDSs are quantified using a traditional injection internal standard approach. The IDS recovery data provide a useful estimate of method performance that is analogous to surrogate recovery performance data provided for other NWQL organic contaminant methods. Analyte method recoveries are automatically corrected for incomplete IDS recovery by using the isotope-dilution quantification procedure, so no additional correction to analyte concentration or recoveries is needed or appropriate.

Comparisons of analyte method recoveries relative to IDS absolute recoveries provide an assessment of the application of the IDQ procedure in this method when using both exact and non-exact isotopic analogs (table 7). Plots of relations between analyte method recoveries and IDS absolute recoveries for the four validation matrices described in this section are shown in figure 8. In these plots, recovery data are not included for analytes having a mean ambient concentration in the matrix that exceeded 100 percent of the fortification concentration.

Reagent Water

The reagent-water validation matrix was from the Solution 2000 water-purification system and contained no detectable steroid hormones. Routine monitoring of the water produced by this system showed that the dissolved organic carbon (DOC) concentration typically was less than 0.016 milligrams per liter (mg/L). Each of nine reagent-water replicates were fortified at the low and high levels (tables 12 and 13, respectively).

Mean IDS recoveries in the reagent-water validation matrix ranged from 67 to 93 percent, with RSDs < 7 percent. Note: In previous method performance testing (not presented in this report), cholesterol-d_7 exhibited very poor recovery (generally <20 percent) in all reagent-water matrices (laboratory and field-submitted reagent-water samples), whereas its recovery in field-water matrices typically was >50 percent (see "Cholesterol-d_7 Recoveries in Non-salted Reagent Water" section). Cholesterol-d_7 recoveries in reagent-water matrices were dramatically improved by the addition of salt to all water samples before extraction, including all sample data described in this report.

Mean method recoveries of the analytes in the reagent-water validation replicates ranged from 78 to 120 percent, with RSDs ≤12 percent for all analytes except equilin, which had an RSD of 22 percent in high-level spikes. These recoveries are within the target method performance range of 60–120 percent mean recovery and ≤25 percent RSD (NWQL SOP MX0015.2, "Guidelines for Method Validation and Publication" (Foreman and Green, U.S. Geological Survey, written commun., 2005)).

A comparison between analyte and IDS recoveries for these reagent-water validation replicates illustrates the expected differences between *absolute* IDS recovery and the corresponding analyte's *method* recovery obtained by using the IDQ procedure (fig. 8). At least theoretically, each analyte's method recovery always will be greater than the IDS's absolute recovery (for an exact isotopic IDS) and is expected to be near 100 percent if the analyte and corresponding IDS emulate each other in absolute recovery during sample preparation and analysis. For example, the mean *absolute* recovery of mestranol-d_4 in the low-level reagent-water spikes was 74 percent, whereas the mean *method* recovery for mestranol analyte was 98 percent. These expected recovery differences between analytes and IDS compounds warrant consideration by data users when interpreting method performance data. As shown in figure 8, all analyte method recoveries fell within 60–120 percent for all reagent-water validation replicates, except for equilin in one replicate that was biased high, and bisphenol A in six low-level replicates that ranged between 120 and 130 percent. The high bias in BPA recovery in the low-level replicates is likely from unaccounted for contamination introduced during sample preparation for this blank-limited compound (see "Blank-Limited Analytes" section).

Surface Water

Replicate samples of the surface-water matrix used for method validation were collected on April 14, 2010, from Rapid Creek about 50 meters downstream from a WWTP outfall near Rapid City, S. Dak. The water had a pH of 8.16 and specific conductance of 930 microsiemens per centimeter at 25°C; concentrations of total suspended solids and DOC were not determined. Mean IDS recoveries in the surface-water matrix spikes ranged from 46 to 88 percent, with RSDs <12 percent, except for medroxyprogesterone-d_3 that had mean recoveries <32 percent and RSDs as high as 44 percent and diethylstilbestrol-d_8 that had mean recoveries <24 percent (RSDs <7 percent) (tables 12–14). Indeed for these two IDSs plus bisphenol A-d_{16} and nandrolone-d_3, the recovery bias in the surface-water matrix was distinctly low.

Mean recoveries for most analytes ranged from 71 to 144 percent, with RSDs <17 percent. Recoveries for cholesterol and 3β-coprostanol were more variable in the low-level spikes because of high mean ambient concentrations (table 14) that were >1,400 percent of the fortification level of 1,000 ng/L. Progesterone was poorly recovered (<12 percent mean) in the surface-water matrix (tables 12 and 13); progesterone's loss in this matrix was substantially greater than that for its corresponding IDS, medroxyprogesterone-d_3 (fig. 8T), indicating that the isotope-dilution quantification procedure using this non-exact IDS analog did not adequately compensate for the amount of progesterone loss in this matrix. This is one example of the limitation of the IDQ procedure when non-exact isotopic analogs are used; the absolute recovery of the analyte is not well emulated by its corresponding IDS. Nevertheless, determined progesterone concentrations in these surface-water matrix spikes are less negatively biased than if the IDQ procedure had not been used. Medroxyprogesterone-d_3 and especially progesterone had poor recoveries in some matrices, the cause for which has not been elucidated. In addition, the half-life of the di-(trimethylsilyl)-derivative of progesterone was determined to be about 3.5 days at ambient temperature (data not shown); although not specifically characterized, the half-life is substantially longer when the extract is stored at <–15°C. Based on these characteristics and previous matrix-spike performance tests not presented in this report, all concentrations for progesterone are reported to NWIS as estimated.

Although progesterone concentrations might be biased low in some matrices, it has been retained as an analyte in this method because of evidence that progesterone might have environmental effects at concentrations substantially lower than those for some other method analytes (Kolodziej and others, 2003), and because it has been consistently detected using this method in influent or primary effluent samples collected from WWTPs. Improvements in matrix-specific method recoveries for progesterone are expected by the recent substitution of the exact isotopic analog progesterone-$^{13}C_3$ for medroxyprogesterone-d_3 (see section 10.7).

Table 12. Bias and variability of the method analytes fortified at low levels in replicate samples of reagent water, surface water, secondary wastewater effluent, and primary wastewater effluent.

[N, number of replicates; NR, not reported because ambient concentrations exceeded 300 percent of fortified amount, which produced substantially skewed recoveries; RSD, relative standard deviation. Some values might have additional bias due to concentrations in the ambient sample from 25 to 150 percent (**bold** values), from 151 to 300 percent (***bold italicized*** values), or greater than 300 percent (***bold italicized*** values with footnote) of the fortified amount. Fortification level was 10 nanograms per liter (ng/L) for 17 analytes, 100 ng/L for bisphenol A, 320 or 1,000 ng/L for 3-*beta*-coprostanol, and 1,000 ng/L for cholesterol, assuming a 0.5-liter sample volume. Isotope-dilution standards were fortified at 100 ng/L, except cholesterol-d_7, which was fortified at 10,000 ng/L]

Analyte	Reagent water N = 9		Surface water N = 8		Secondary wastewater effluent N = 9		Primary wastewater effluent N = 8	
	Mean recovery (percent)	RSD (percent)	Mean recovery (percent)	RSD (percent)	Mean recovery (percent)	RSD (percent)	Mean recovery (percent)	RSD (percent)
Method analytes								
11-Ketotestosterone	100	3.2	83.4	16.3	106	2.9	*145*[a]	*33.1*[a]
17-*alpha*-Estradiol	103	1.6	103	7.8	107	5.5	149	8.3
17-*alpha*-Ethynylestradiol	98.7	3.5	82.2	9.3	92.8	1.6	86.4	13.0
17-*beta*-Estradiol	102	1.9	92.3	8.4	103	4.4	**94.3**	**10.8**
3-*beta*-Coprostanol	95.7	2.9	*113*[a]	*65.1*[a]	**89.9**	**21.0**	NR	NR
Androstenedione	98.3	2.7	109	9.9	97.3	4.4	NR	NR
Bisphenol A	120	6.4	72.5	8.0	**171**[b]	**5.7**	NR	NR
Cholesterol	83.1	2.3	*174*[a]	*39.9*[a]	88.1	15.1	NR	NR
cis-Androsterone	96.6	2.7	118	8.7	80.2	4.9	NR	NR
Dihydrotestosterone	98.0	4.4	107	13.9	96.0	6.5	*85.9*[a]	*102*[a]
Epitestosterone	97.5	1.8	121	6.7	104	3.1	*126*[a]	*27.9*[a]
Equilenin	93.5	5.7	92.6	11.6	82.8	6.7	63.8	18.1
Equilin	91.3	9.5	107	9.1	287[c]	10.0	*120*[a]	*84.5*[a]
Estriol	88.4	3.1	75.1	11.0	88.3	2.9	*75.5*[a]	*224*[a]
Estrone	103	2.2	**93.5**	**8.1**	103	4.8	*110*[a]	*38.4*[a]
Mestranol	97.8	3.6	93.3	7.0	101.7	4.2	82.3	9.3
Norethindrone	98.3	3.7	99.4	9.2	95.5	4.2	97.4	9.1
Progesterone	90.8	3.2	8.9	25.3	75.4	5.2	NR	NR
Testosterone	97.5	4.1	104	8.7	94.0	4.4	NR	NR
trans-Diethylstilbestrol	95.2	3.5	76.1	9.2	93.3	2.5	88.1	6.9
Isotope-dilution standards								
16-Epiestriol-d_2	78.3	6.0	88.4	6.3	80.1	4.7	87.1	11.5
17-*alpha*-Ethynylestradiol-d_4	80.8	4.4	77.8	4.1	76.1	4.0	66.5	10.8
17-*beta*-Estradiol-$^{13}C_6$	80.1	3.6	66.1	4.5	72.0	4.1	57.5	10.8
Bisphenol A-d_{16}	81.5	4.4	58.9	6.2	52.9	6.5	65.9	22.6
Cholesterol-d_7	78.6	3.8	71.6	6.1	72.9	13.7	44.4	9.1
Diethylstilbestrol-d_8	67.3	5.7	19.8	5.2	55.2	5.7	83.6	10.8
Estrone-$^{13}C_6$	79.6	4.9	77.5	5.6	78.6	3.5	65.1	9.9
Medroxyprogesterone-d_3	78.4	4.8	26.5	38.7	110	6.3	82.8	8.1
Mestranol-d_4	74.5	4.5	70.7	3.5	70.3	4.7	67.4	7.4
Nandrolone-d_3	81.4	4.5	55.0	7.8	85.0	3.7	61.2	10.6

[a]Values provided to show recovery and RSD even when the ambient or interference concentration exceeded 300 percent of amount fortified.

[b]High bias in the bisphenol A recovery likely due to fortification of bisphenol A-d_{16} at 100-times normal level in error for the secondary wastewater effluent.

[c]Equilin recovery in the secondary wastewater effluent was unexpectedly high for unknown reasons; no interference was noted.

Table 13. Bias and variability of the method analytes fortified at high levels in replicate samples of reagent water, surface water, secondary wastewater effluent, and primary wastewater effluent.

[N, number of replicates; NR, not reported because ambient concentrations exceeded 300 percent of fortified amount producing skewed recoveries; RSD, relative standard deviation. Some values might have additional bias due to concentrations in the ambient sample between 25 and 150 percent (**bold** values) or 150 to 300 percent (***bold italicized values***) of the fortified amount. Fortification level was 100 nanograms per liter (ng/L) for 17 analytes, 1,000 ng/L for bisphenol A, 3,200 or 10,000 ng/L for 3-*beta*-coprostanol, and 10,000 ng/L for cholesterol assuming a 0.5-liter nominal sample volume. Isotope-dilution standards were fortified at 100 ng/L, except cholesterol-d_7, which was fortified at 10,000 ng/L.]

Analyte	Reagent water N= 9		Surface water N = 8		Secondary wastewater effluent N = 8		Primary wastewater effluent N = 8	
	Mean recovery (percent)	RSD (percent)	Mean recovery (percent)	RSD (percent)	Mean recovery (percent)	RSD (percent)	Mean recovery (percent)	RSD (percent)
Method analytes								
11-Ketotestosterone	104	6.5	70.6	11.9	117	6.8	**87.7**	**6.1**
17-*alpha*-Estradiol	101	7.1	104	5.8	114	3.7	138	5.0
17-*alpha*-Ethynylestradiol	97.8	4.8	81.5	2.3	94.8	3.3	97.8	3.0
17-*beta*-Estradiol	104	6.9	94.4	6.0	108	4.6	94.6	4.4
3-*beta*-Coprostanol	107	6.2	82.3	6.7	141	6.4	NR	NR
Androstenedione	100	5.9	124	5.3	106	2.3	NR	NR
Bisphenol A	98.1	6.2	82.8	3.5	168[a]	13.6	**101**	**7.9**
Cholesterol	102	5.7	78.8	4.1	108	5.9	NR	NR
cis-Androsterone	91.2	5.7	144	8.9	96.6	3.2	NR	NR
Dihydrotestosterone	94.9	5.3	116	8.6	106	2.5	***88.1***	***9.9***
Epitestosterone	96.4	5.5	131	6.4	109	2.1	110	2.6
Equilenin	78.4	12.0	92.8	4.6	82.4	7.1	103	6.7
Equilin	118	22.4	122	3.3	222[b]	8.9	**147**	**11.7**
Estriol	92.4	4.8	79.9	3.3	89.3	2.9	***106***	***29.7***
Estrone	100	4.1	93.2	3.9	105	4.0	**116**	**8.0**
Mestranol	98.8	5.8	92.1	4.3	104	3.8	92.9	3.1
Norethindrone	103	5.4	96.6	5.2	99.6	5.4	102	2.7
Progesterone	88.8	6.7	11.9	40.0	78.6	4.9	**45.9**	**35.6**
Testosterone	99.2	5.7	122	7.1	98.8	3.1	***103***	***9.2***
trans-Diethylstilbestrol	92.4	4.6	74.7	2.6	98.7	3.0	95.1	3.2
Isotope-dilution standards								
16-Epiestriol-d_2	88.5	5.5	87.7	4.2	77.2	6.7	88.0	7.8
17-*alpha*-Ethynylestradiol-d_4	89.4	3.4	79.1	2.3	72.0	4.3	59.1	3.9
17-*beta*-Estradiol-$^{13}C_6$	88.4	5.2	67.4	5.2	69.7	3.3	51.8	5.9
Bisphenol A-d_{16}	93.0	6.2	58.5	5.1	56.5	13.9	79.6	4.8
Cholesterol-d_7	70.0	5.6	87.5	6.4	62.7	10.8	41.3	14.0
Diethylstilbestrol-d_8	76.6	3.8	21.9	6.5	55.9	10.0	73.5	4.7
Estrone-$^{13}C_6$	90.0	3.6	78.9	4.9	78.2	3.3	61.5	4.8
Medroxyprogesterone-d_3	91.5	8.0	15.6	43.7	104	11.0	84.8	6.8
Mestranol-d_4	79.8	5.2	71.6	3.6	67.1	4.0	59.2	3.9
Nandrolone-d_3	86.2	2.6	45.9	11.8	81.3	3.5	60.0	4.2

[a]High bias in the bisphenol A recovery likely due to fortification of bisphenol A-d_{16} at 100-times normal level in error for the secondary wastewater effluent.

[b]Equilin recovery in the secondary wastewater effluent was unexpectedly high for unknown reasons; no interference was noted.

Table 14. Bias and variability of method analyte concentrations and isotope-dilution standard recoveries in unspiked, quadruplicate samples of validation matrices from surface water, secondary wastewater effluent, and primary wastewater effluent.

[NA, not applicable; --, not detected; ng/L, nanogram per liter; RSD, relative standard deviation; <, less than. Isotope-dilution standards were fortified at 100 ng/L, except cholesterol-d_7, which was fortified at 10,000 ng/L]

Analyte	Surface water		Secondary wastewater effluent		Primary wastewater effluent	
	Mean concentration (ng/L)	RSD (percent)	Mean concentration (ng/L)	RSD (percent)	Mean concentration (ng/L)	RSD (percent)
Method analytes						
11-Ketotestosterone	--	NA	--	NA	40.7	7.9
17-*alpha*-Estradiol	0.1	18.2	--	NA	--	NA
17-*alpha*-Ethynylestradiol	--	NA	--	NA	--	NA
17-*beta*-Estradiol	0.6	18.2	--	NA	9.4	7.1
3-*beta*-Coprostanol	13,920	7.8	147[a]	69.9	816,100[b]	15.1
Androstenedione	2.0	18.6	1.2[c]	15.3	424[b]	12.6
Bisphenol A	14.7[a]	1.5	25.8[a,d]	7.9	705	11.2
Cholesterol	17,060	6.7	113[a]	14.5	1,249,400[b]	10.6
cis-Androsterone	1.9	14.8	<2.4[c]	1.6	2,315[b]	27.1
Dihydrotestosterone	--	NA	--	NA	127	4.8
Epitestosterone	--	NA	--	NA	47.7	3.0
Equilenin	--	NA	--	NA	--	NA
Equilin	2.6[c]	15.0	--	NA	<57[c]	10.6
Estriol	1.9	14.0	--	NA	234	4.4
Estrone	4.8	9.4	--	NA	54.9	7.2
Mestranol	--	NA	--	NA	--	NA
Norethindrone	--	NA	--	NA	0.6	21.3
Progesterone	--	NA	--	NA	36	25.6
Testosterone	0.5[c]	NA	--	NA	171	3.6
trans-Diethylstilbestrol	0.5[c]	19.9	--	NA	--	NA

Analyte	Mean recovery (percent)	RSD (percent)	Mean recovery (percent)	RSD (percent)	Mean recovery (percent)	RSD (percent)
Isotope-dilution standards						
16-Epiestriol-d_2	94.4	4.0	85.2	11.1	99.3	1.8
17-*alpha*-Ethynylestradiol-d_4	82.0	3.4	91.4	16.7	82.0	7.7
17-*beta*-Estradiol-$^{13}C_6$	69.8	3.7	92.6	21.6	71.3	8.4
Bisphenol A-d_{16}	61.9	1.4	87.6[d]	20.1	88.4	8.2
Cholesterol-d_7	77.0	8.5	52.2	12.3	48.5	9.3
Diethylstilbestrol-d_8	23.9	5.3	74.9	22.4	99.1	5.9
Estrone-$^{13}C_6$	82.1	3.9	101	18.1	79.2	12.3
Medroxyprogesterone-d_3	31.2	21.2	101	5.0	89.5	2.1
Mestranol-d_4	74.5	4.2	85.2	17.4	81.0	8.2
Nandrolone-d_3	59.1	6.8	101	16.4	79.8	8.1

[a]The mean ambient concentration was used for correction of matrix spike recoveries even though its value is less than the analyte's minimum reporting level.

[b]Extract dilution was required to quantify the analyte in replicates from the primary wastewater effluent.

[c]Mean unspiked concentration shown was used for background correction of spike recoveries, but the concentration in one or more of the unspiked replicates was not reported because mass spectral qualification criteria were not met to confirm analyte presence.

[d]Bisphenol A-d_{16} was fortified at 100-times normal level in error for one replicate of the secondary wastewater effluent. The high-biased bisphenol A and bisphenol A-d_{16} values from that replicate were omitted from the calculation of the mean.

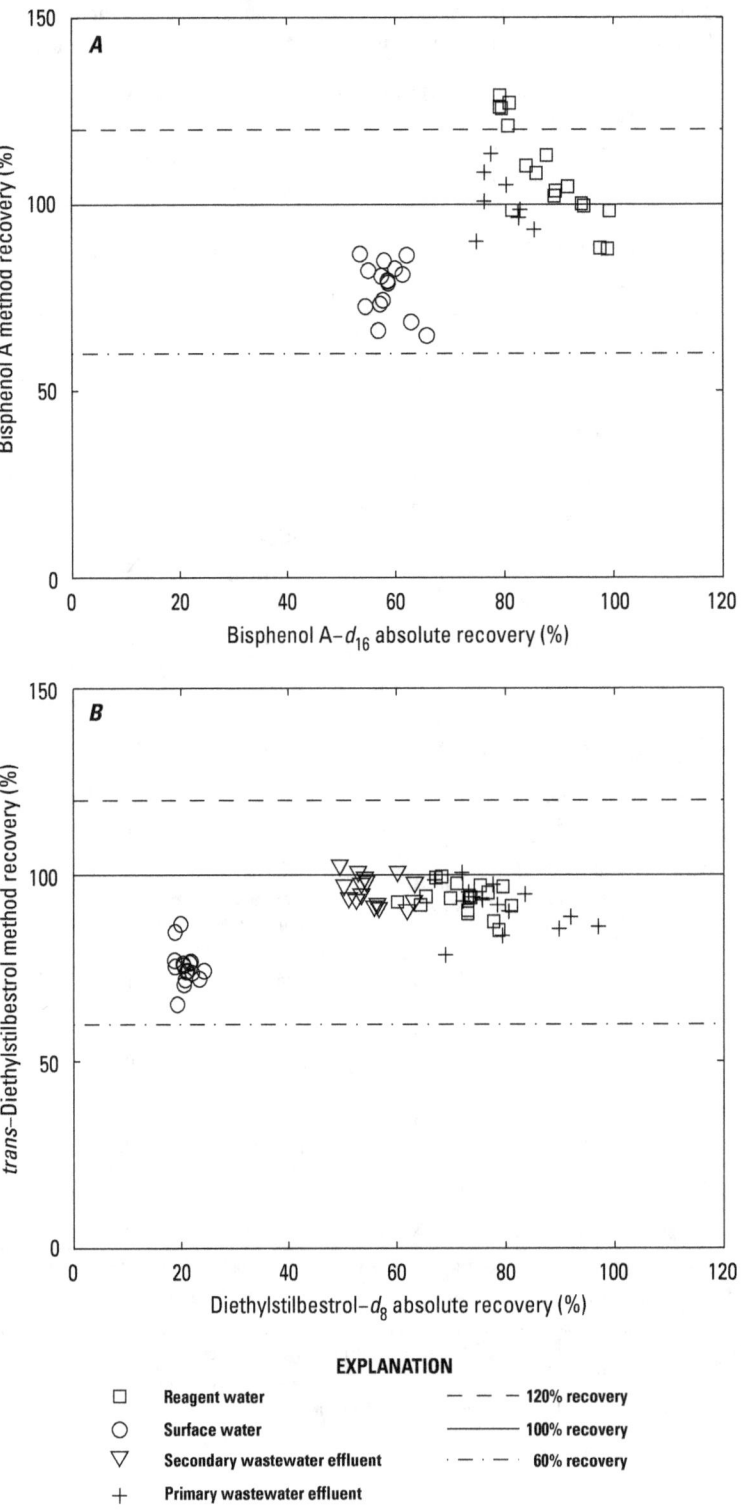

Figure 8.   Relation between analyte method recoveries and isotope-dilution standard absolute recoveries in percent (%) in spiked replicates of reagent water, surface water, secondary wastewater effluent, and primary wastewater effluent validation matrices. Samples with mean ambient analyte concentrations that exceeded the fortified concentrations were excluded to eliminate potential bias.

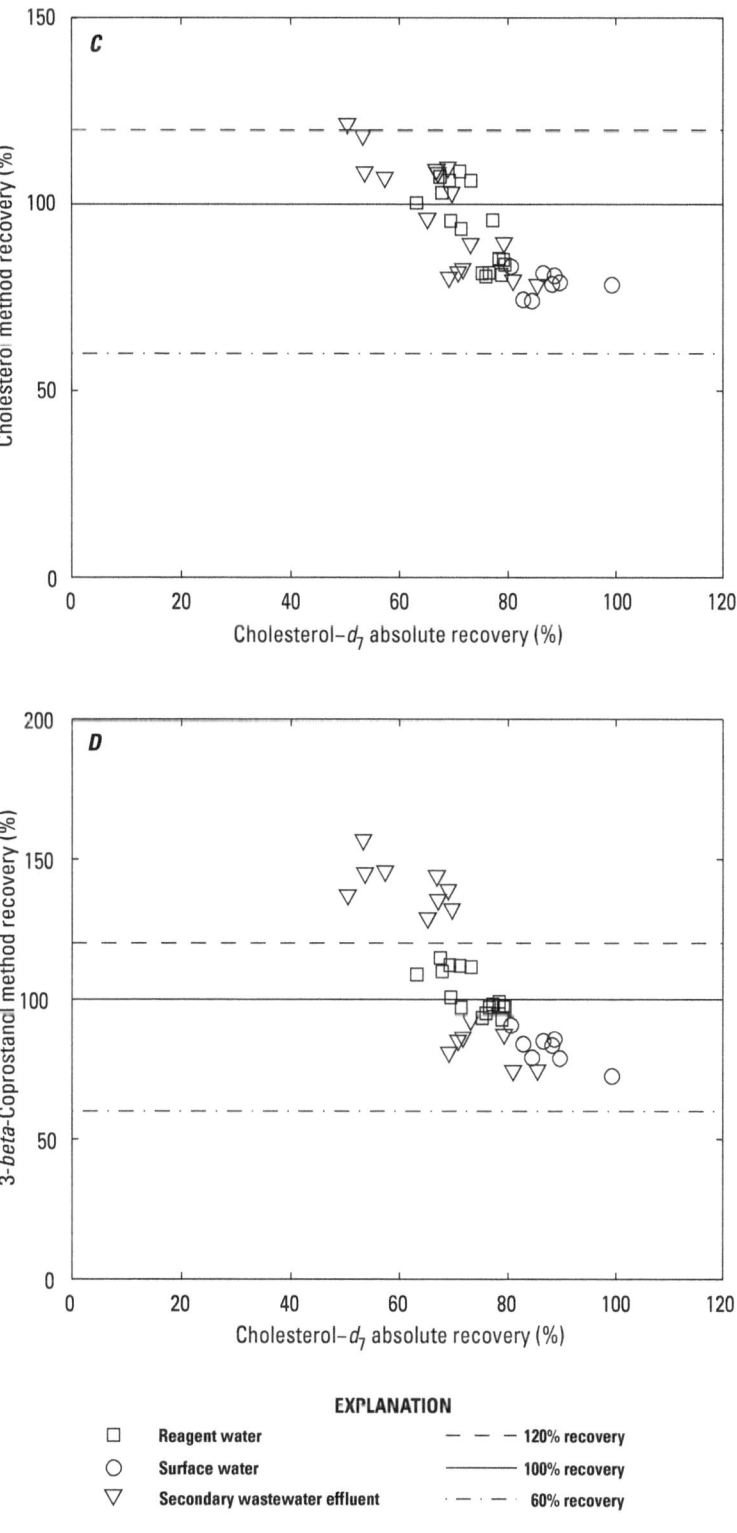

Figure 8. Relation between analyte method recoveries and isotope-dilution standard absolute recoveries in percent (%) in spiked replicates of reagent water, surface water, secondary wastewater effluent, and primary wastewater effluent validation matrices. Samples with mean ambient analyte concentrations that exceeded the fortified concentrations were excluded to eliminate potential bias.—Continued

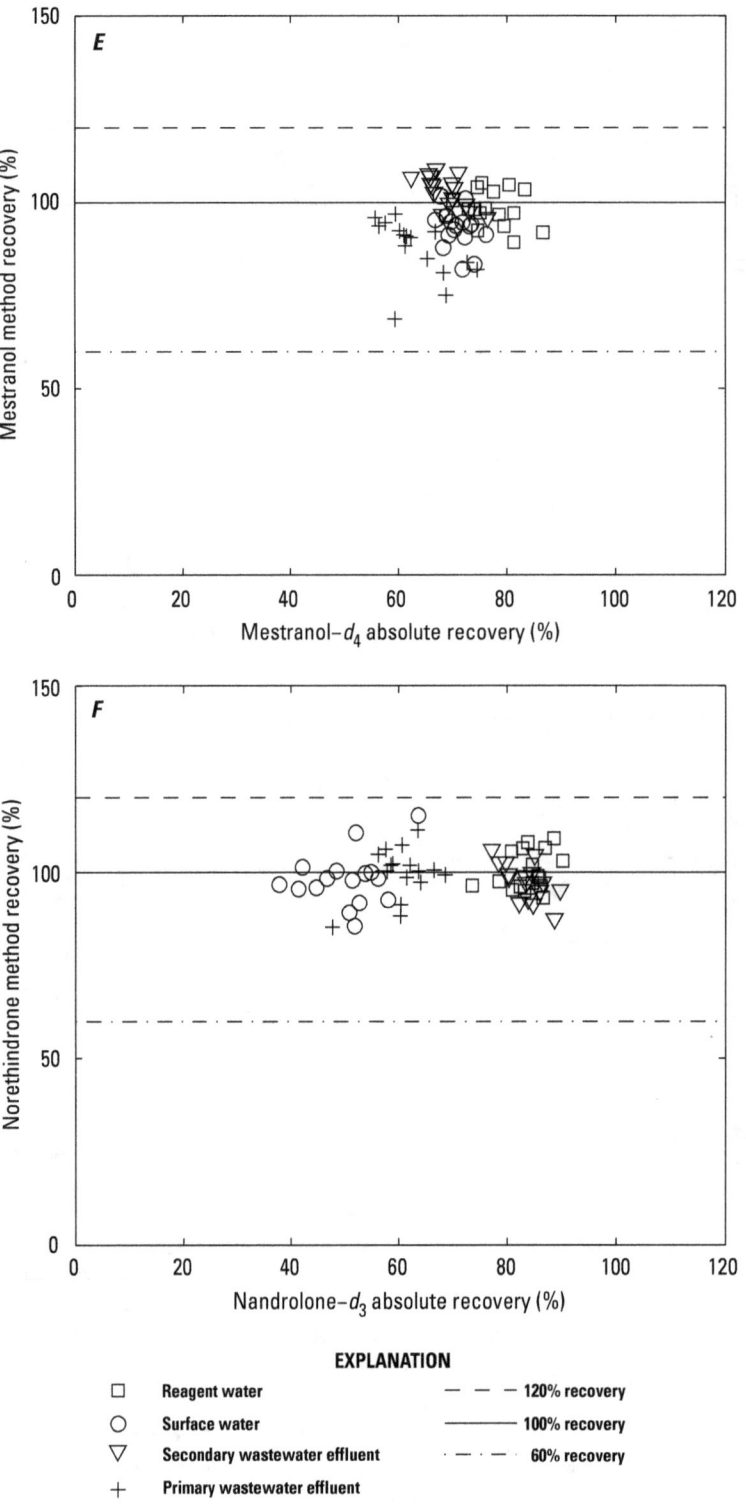

Figure 8. Relation between analyte method recoveries and isotope-dilution standard absolute recoveries in percent (%) in spiked replicates of reagent water, surface water, secondary wastewater effluent, and primary wastewater effluent validation matrices. Samples with mean ambient analyte concentrations that exceeded the fortified concentrations were excluded to eliminate potential bias.—Continued

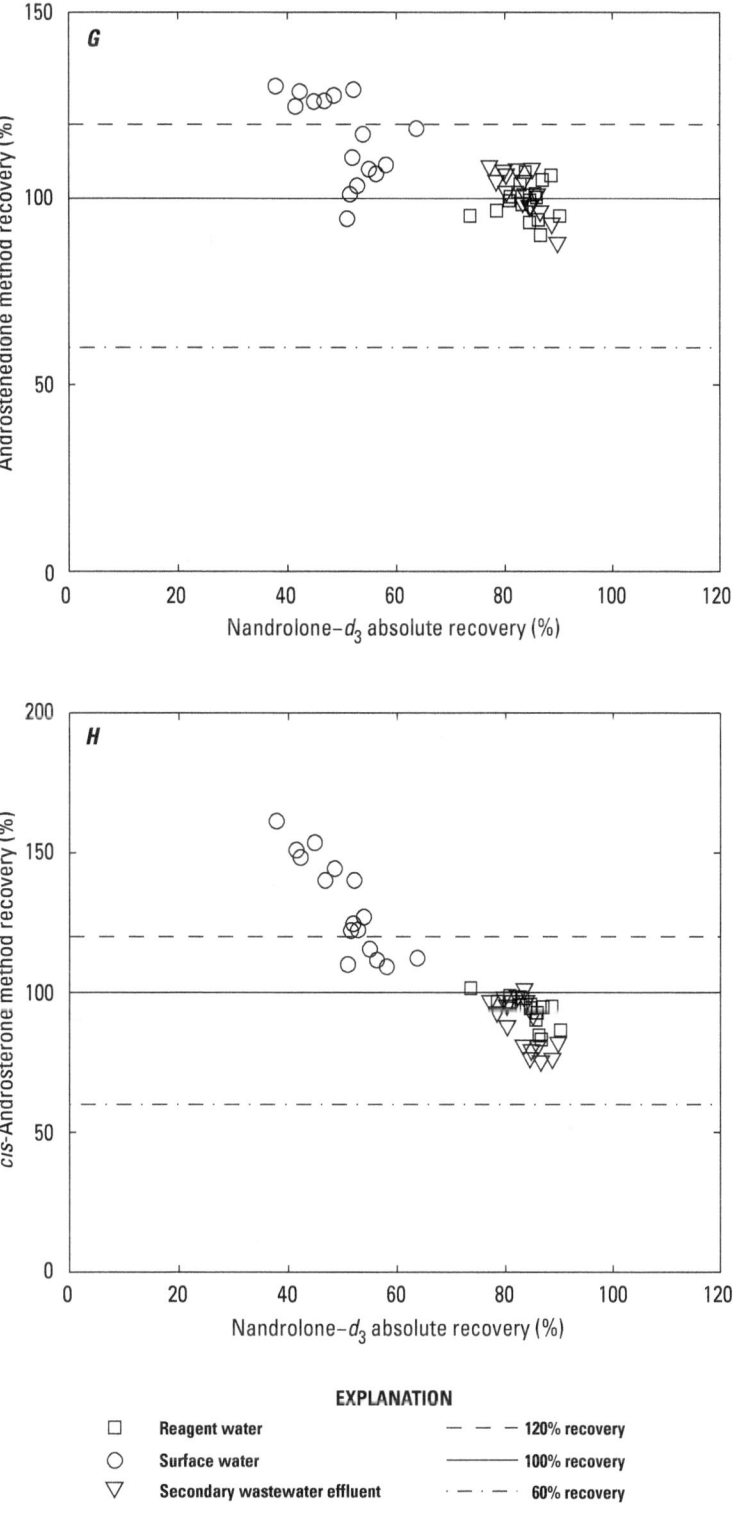

Figure 8. Relation between analyte method recoveries and isotope-dilution standard absolute recoveries in percent (%) in spiked replicates of reagent water, surface water, secondary wastewater effluent, and primary wastewater effluent validation matrices. Samples with mean ambient analyte concentrations that exceeded the fortified concentrations were excluded to eliminate potential bias.—Continued

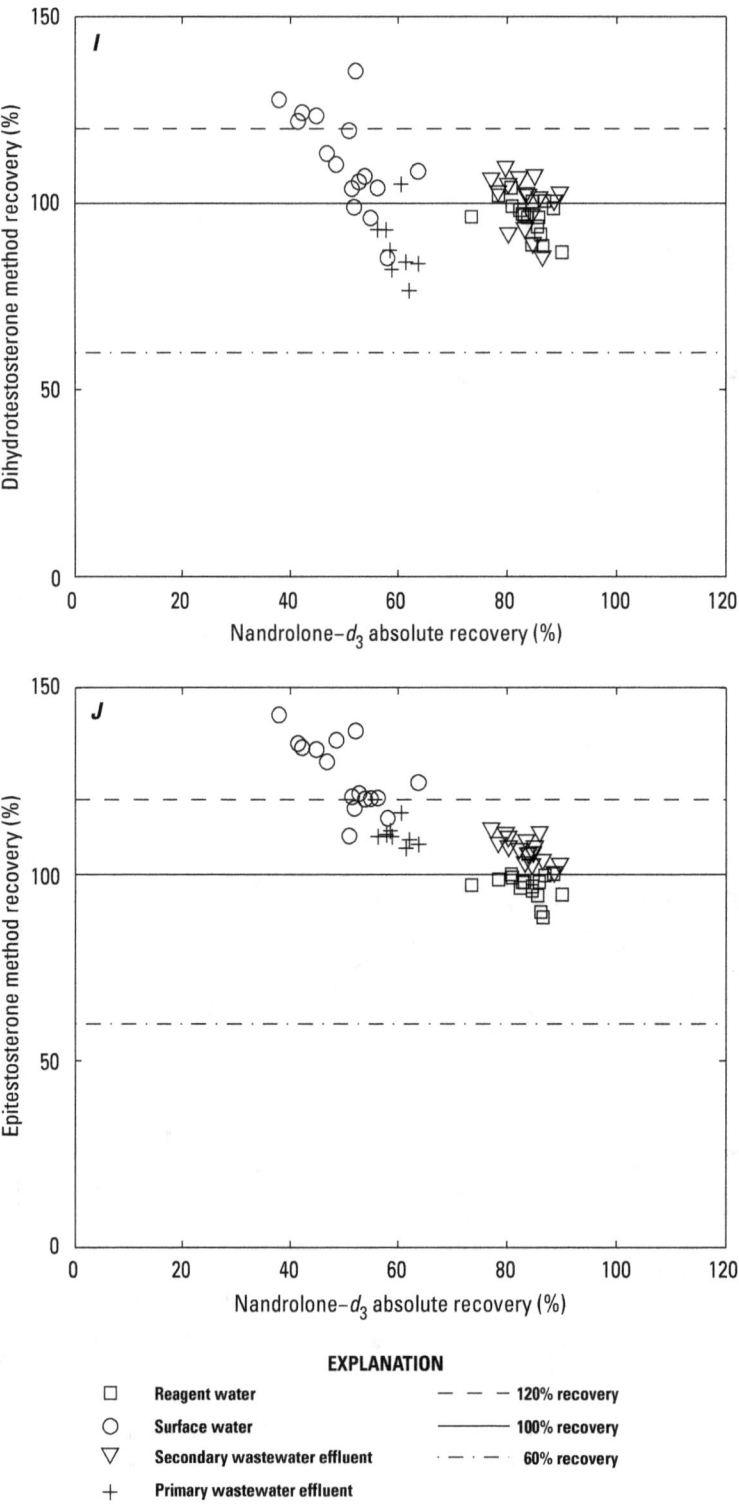

Figure 8. Relation between analyte method recoveries and isotope-dilution standard absolute recoveries in percent (%) in spiked replicates of reagent water, surface water, secondary wastewater effluent, and primary wastewater effluent validation matrices. Samples with mean ambient analyte concentrations that exceeded the fortified concentrations were excluded to eliminate potential bias.—Continued

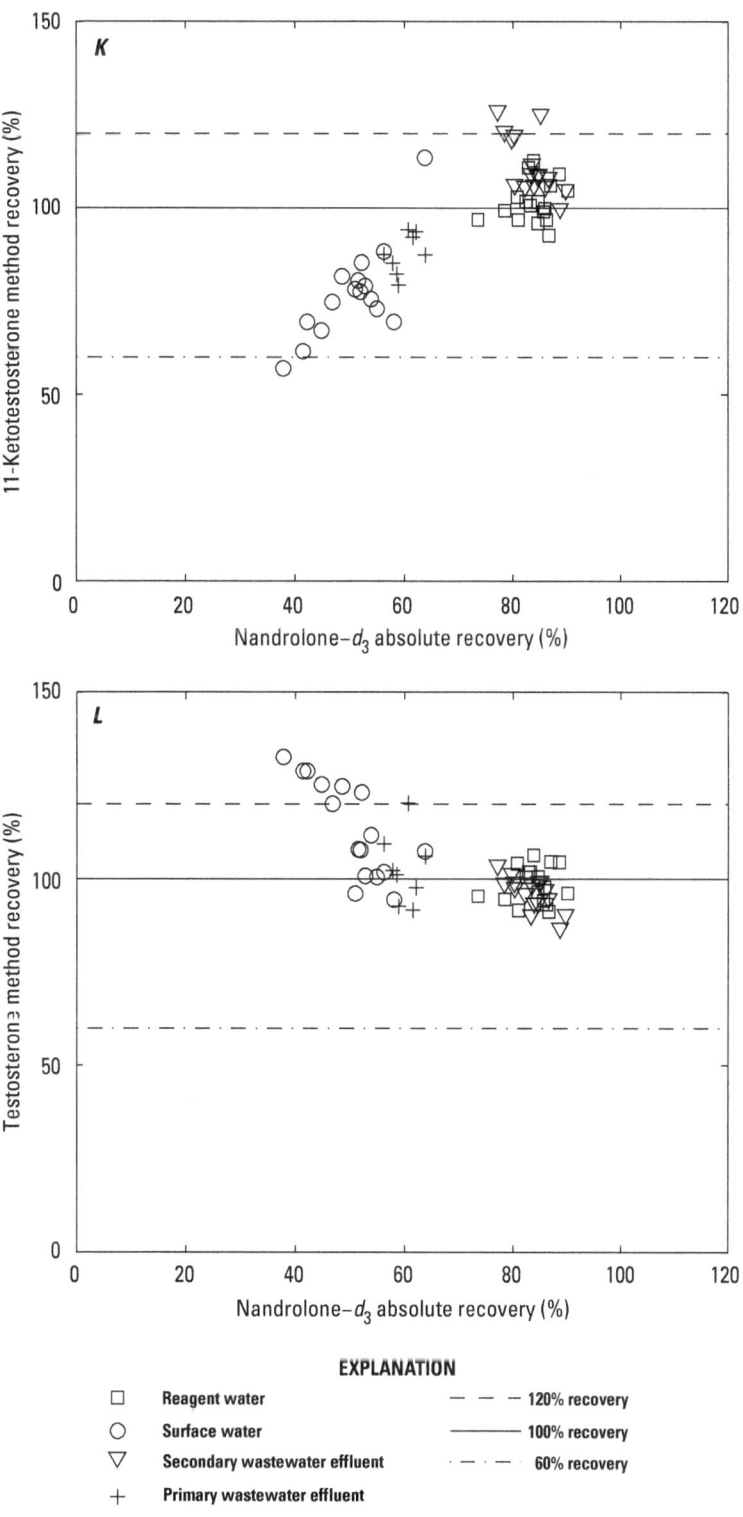

Figure 8. Relation between analyte method recoveries and isotope-dilution standard absolute recoveries in percent (%) in spiked replicates of reagent water, surface water, secondary wastewater effluent, and primary wastewater effluent validation matrices. Samples with mean ambient analyte concentrations that exceeded the fortified concentrations were excluded to eliminate potential bias.—Continued

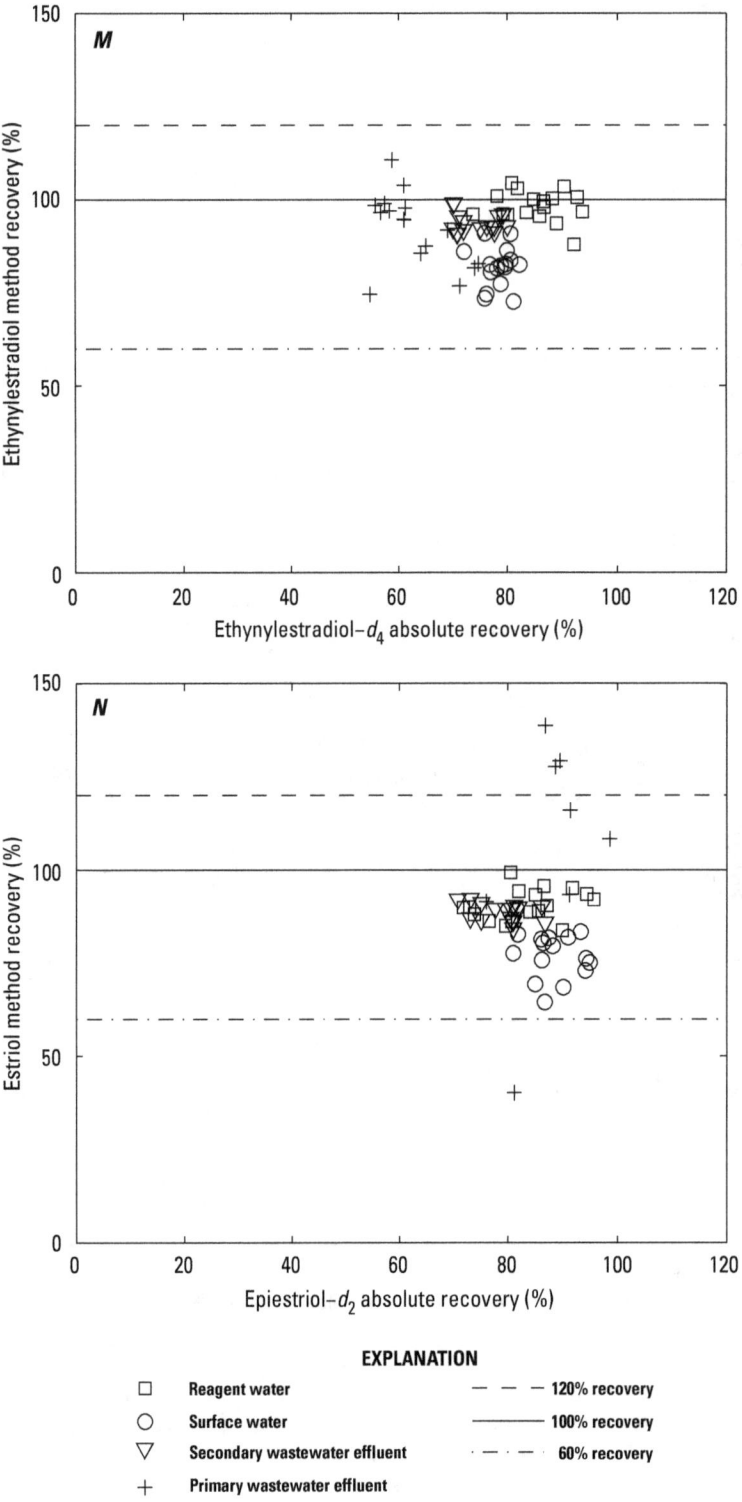

Figure 8.   Relation between analyte method recoveries and isotope-dilution standard absolute recoveries in percent (%) in spiked replicates of reagent water, surface water, secondary wastewater effluent, and primary wastewater effluent validation matrices. Samples with mean ambient analyte concentrations that exceeded the fortified concentrations were excluded to eliminate potential bias.—Continued

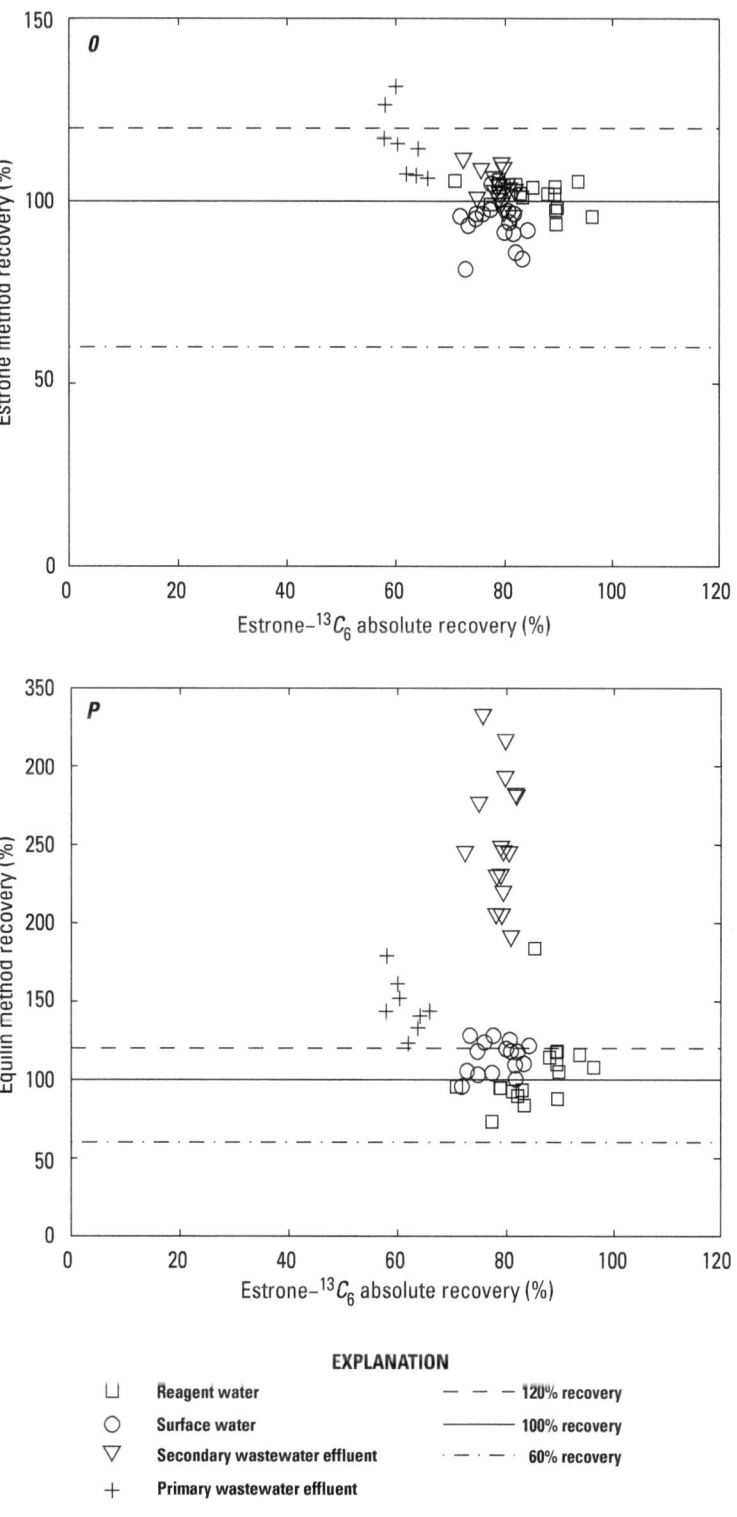

Figure 8. Relation between analyte method recoveries and isotope-dilution standard absolute recoveries in percent (%) in spiked replicates of reagent water, surface water, secondary wastewater effluent, and primary wastewater effluent validation matrices. Samples with mean ambient analyte concentrations that exceeded the fortified concentrations were excluded to eliminate potential bias.—Continued

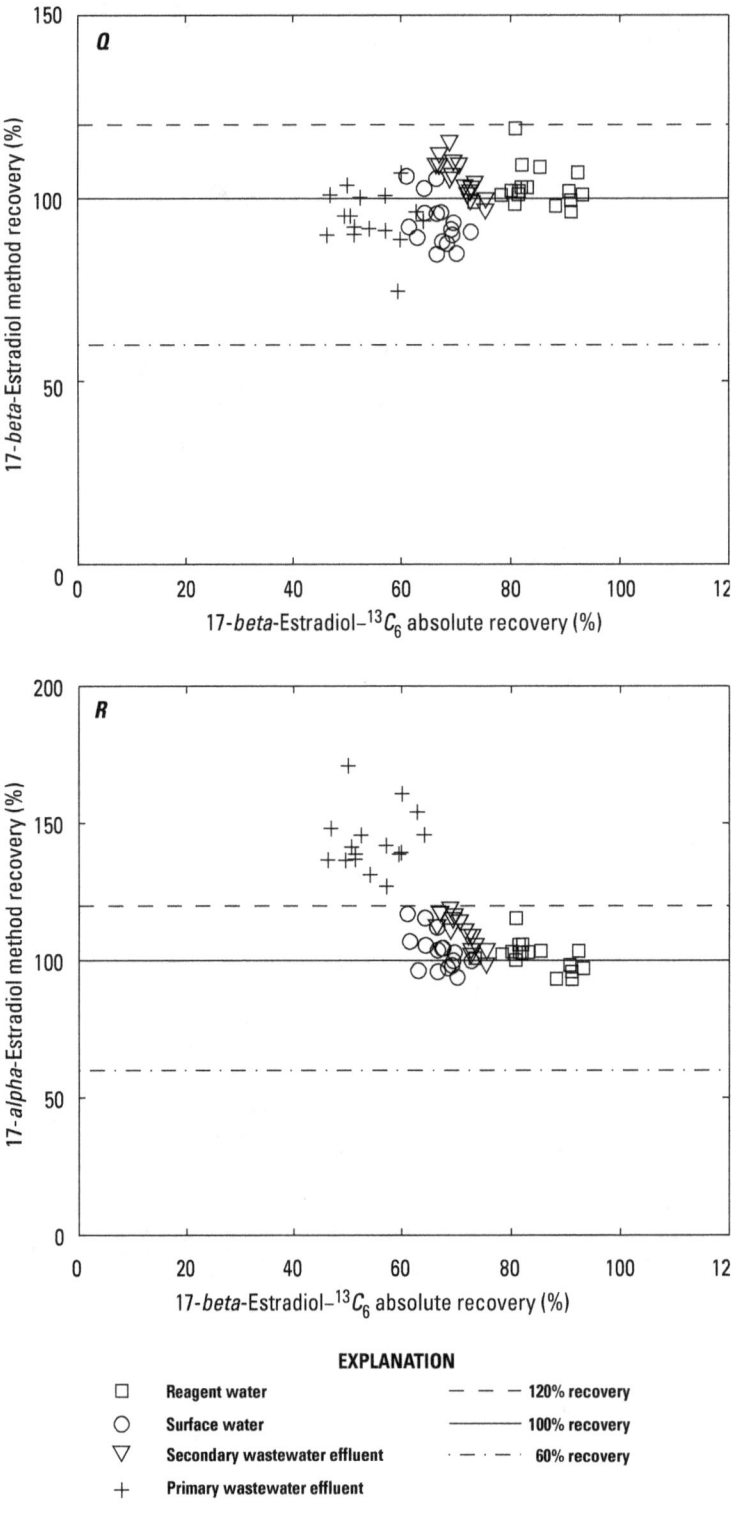

Figure 8. Relation between analyte method recoveries and isotope-dilution standard absolute recoveries in percent (%) in spiked replicates of reagent water, surface water, secondary wastewater effluent, and primary wastewater effluent validation matrices. Samples with mean ambient analyte concentrations that exceeded the fortified concentrations were excluded to eliminate potential bias.—Continued

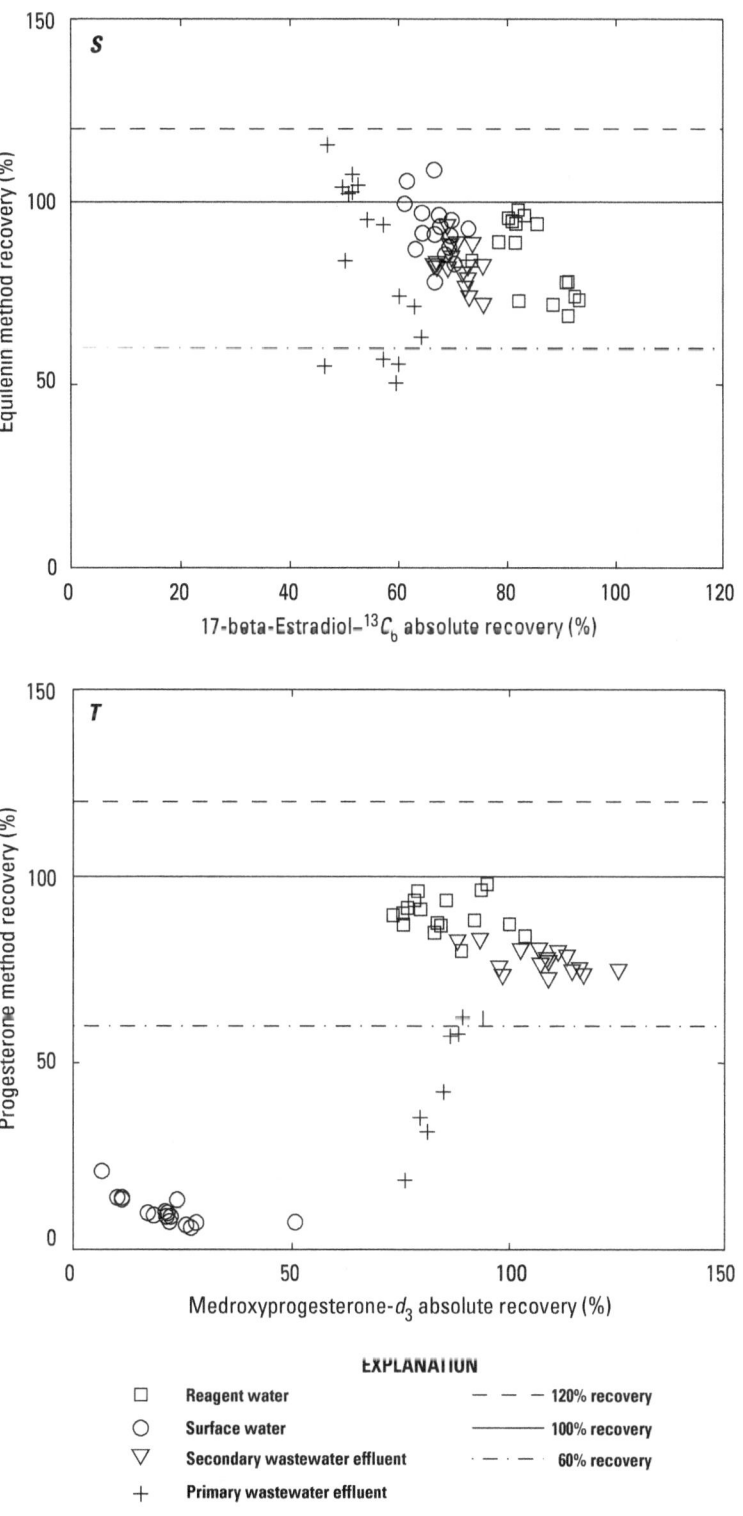

Figure 8.   Relation between analyte method recoveries and isotope-dilution standard absolute recoveries in percent (%) in spiked replicates of reagent water, surface water, secondary wastewater effluent, and primary wastewater effluent validation matrices. Samples with mean ambient analyte concentrations that exceeded the fortified concentrations were excluded to eliminate potential bias.—Continued

The mean recovery of diethylstilbestrol (75 percent) in the surface-water matrix is lower than in the other validation matrices (tables 12 and 13; fig. 8*B*), but is well within the target performance range of 60–120 percent mean recovery and demonstrates the advantage and applicability of the IDQ procedure even when the IDS recovery is low (in this case, about 20 percent recovery for diethylstilbestrol-d_8 at both spiking levels).

Secondary Wastewater Effluent

Replicates samples of the secondary wastewater effluent matrix used during method validation were collected on March 11, 2010, from a WWTP in New York (identified as NY3 by Phillips and others, 2010). The water had a pH of 6.9 (data provided by the plant operator on a separate aliquot); concentrations of DOC and total suspended solids were not determined, although total suspended solids from effluents samples collected monthly by the plant operator are normally <4 mg/L. Use of ascorbic acid was not necessary for this matrix because ultraviolet treatment is used instead of chlorination for disinfection by this WWTP. Nine replicate samples were spiked at the low level and eight replicate samples were spiked at the high level. Mean IDS recoveries in these secondary wastewater effluent matrix spikes ranged from 53 to 110 percent, with RSDs <14 percent (tables 12 and 13). RSDs for most of the IDS compounds in the unspiked (ambient) replicates were somewhat higher than in the spiked replicates but were still <23 percent (table 14).

Mean recoveries for most analytes ranged from 75 to 117 percent, with RSDs <15 percent. Bisphenol A-d_{16} was spiked in error at 100 times the normal fortification level in both the low and high spikes. This error probably produced the high biased recoveries (means of about 170 percent) for bisphenol A, which are not plotted in figure 8*A*. Equilin had unusually high recoveries (>220 percent mean) at both spike levels (tables 12 and 13), which is not readily explained because no interference was apparent from the GC/MS/MS analysis and equilin was not detected in the unspiked ambient replicate samples for this matrix (table 14). The IDS estrone-$^{13}C_6$, which is used to quantify equilin, had highly reproducible recoveries of about 78 percent in these spikes (fig. 8*P*). In previous performance testing in various matrices not presented in this report, equilin was found to have more variable recoveries; all concentrations for equilin are reported to NWIS as estimated.

Primary Wastewater Effluent

Replicate samples of the primary wastewater effluent matrix used for method validation were collected June 29, 2010, from a WWTP in New York (site identified as NY2-I in Phillips and others, 2010). The sampling location was after partial particle removal by sedimentation from the incoming WWTP flow but before any biological or other treatments. (Note: the location used for collection of the primary wastewater effluent matrix is referred to as an "influent" sampling location by Phillips and others (2010)). This was

the most challenging validation matrix examined due to the presence (observational only) of high amounts of dissolved, colloidal, and particulate organic matter. In many cases, the presence of high concentrations of method analytes in the ambient replicates relative to the fortification levels (10 and 100 ng/L) confounded or prevented accurate calculation of analyte recovery. However, use of this type of matrix was considered an important test of method performance because municipal wastewater can be a major source of steroids to the environment, depending on level of treatment and other operational conditions (for example, treatment bypass of wastewater during storm events). Primary effluent contains elevated concentrations of steroid hormones compared to secondary-treated effluent because the particle-removal and biological processes used by many WWTPs as secondary treatment remove a substantial fraction of trace organic compounds, including estrogens and especially androgens and progestins that are present in primary effluent (Furlong and others, 2011). Studies designed to test the efficiency of engineered technologies for removal of the method analytes during the WWTP processes will necessarily examine this type of complex sample matrix; for example, see Liu and others (2009b) and references therein.

Eight replicates each of the primary wastewater effluent matrix were fortified at the low and high levels. Mean IDS recoveries in the spiked primary wastewater effluent replicates ranged from 41 to 88 percent, with RSDs less than 23 percent (tables 12 and 13). The estrogen isotopes 17α-ethynylestradiol-d_4, 17β-estradiol-$^{13}C_6$, and estrone-$^{13}C_6$ had somewhat lower recoveries in this complex matrix compared to the other matrices (fig. 8). High ambient concentrations (table 14) precluded reporting recoveries for bisphenol A, testosterone, and progesterone in the low-level spikes and for 3β-coprostanol, androstenedione, cholesterol, and *cis*-androsterone in the low- and high-level spikes. Mean recoveries for those analytes with ambient concentrations less than 25 percent of the fortification level ranged from 64 to 149 percent (RSDs <19 percent), with only 17α-estradiol (149 percent in low-level spikes, 138 percent in high-level spikes) having mean recoveries outside the target performance range of 60–120 percent (fig. 8).

Also shown in tables 12 and 13 (in bold and bold italics type) are recoveries and RSDs for several analytes that had sizable ambient concentrations. Although recovery variation as described by the RSD was substantially greater than 25 percent for some of these analytes, especially in the low-level spikes, mean recoveries ranged from 46 to 147 percent in this complex matrix and were within the target performance range of 60–120 percent for most of the analytes, demonstrating reasonable method performance at both low- and high-fortification levels in the presence of substantial co-extracted organic material.

Analyte Variability in Unspiked Validation Matrices

Table 14 shows mean ambient concentrations and RSDs of analytes determined in the quadruplicate samples for the three non-reagent-water (unspiked) validation matrices. Concentrations for some analytes were in the low range of the method, whereas others were in the upper range, especially in the primary wastewater effluent matrix. Indeed, 3β-coprostanol, cholesterol, androstenedione, and *cis*-androsterone concentrations were determined using dilutions of the primary wastewater effluent matrix extracts. The RSDs were ≤27 percent for all analytes in these matrices (except 3β-coprostanol in the secondary wastewater effluent matrix; all but one of the ambient secondary wastewater effluent replicates had 3β-coprostanol concentrations that were less than the MRL and would not be reported).

Comparison of Validation Results

Recoveries in all matrices were within the target performance range of 60–120 percent for most analytes (tables 12 and 13; fig. 8). There were significant, albeit small, differences in recovery performance between matrices. For each method analyte, pairwise comparisons were made of the performance for each possible pair of matrices and fortification level based on individual sample recoveries. The nonparametric Wilcoxon rank-sum test (Helsel and Hirsch, 2002) was used to test for statistical significance (p-value <0.05) and, if distributions were significantly different from each other, the matrix with higher recovery is shown in table 15. The IDS compounds were used to evaluate relative performance between matrices because IDS recovery was not biased by ambient concentrations. Reagent water showed the best overall performance relative to the other matrices; out of 60 possible IDS comparisons in paired matrices, recovery was as good or better in reagent water than the other matrix 86 percent of the time (p-value <0.05). The three field matrices were compared to one another, excluding reagent water (40 possible comparisons). The secondary wastewater effluent had the best average IDS recovery (as good or better recovery in 77 percent of these possible comparisons), followed by surface water (70 percent), and the primary wastewater effluent (50 percent).

The pattern of higher reagent-water recoveries compared to the other three matrices was not as consistent for the analytes as for the IDS compounds. This is likely due to two factors. First, certain analytes were present in the ambient samples at levels comparable to or sometimes exceeding the fortification concentrations (table 14), which adds uncertainty to recovery calculations. Second, analyte concentrations are calculated relative to the IDS compounds, so there is a possibility of some positive analyte bias if the performance of the analyte is not exactly emulated by (is better than) the corresponding IDS compound (fig. 8). For example, the somewhat higher mean recoveries for epitestosterone (121 percent in low-level spikes and 131 percent in high-level spikes) in the surface-water matrix indicate that this analyte experienced less absolute mass loss during sample preparation relative to its corresponding non-exact IDS compound, nandrolone-d_3 (tables 12 and 13; fig. 8*J*). Nevertheless,

the IDQ procedure in general provided acceptable method recoveries for most analytes in these four validation matrices.

When all the recovery data are aggregated by validation matrix (table 16), overall mean IDS recovery is highest in reagent water, followed by secondary wastewater effluent, primary wastewater effluent, and surface water: overall mean range of 61–82 percent; overall RSD range of 10–38 percent. The overall mean IDS recovery in the surface-water matrix is biased low by a matrix effect producing low medroxyprogesterone-d_3 and diethylstilbestrol-d_8 recoveries (tables 12 and 13). Again, the magnitude of the observed differences between the four matrices generally was small (tables 12 and 13).

The analyte summaries shown in table 16 omit recovery data for analytes that have mean ambient concentrations in the matrix greater than 300 percent of the fortification concentration. Overall mean recovery of the analytes in the four matrices ranged from 93 to 110 percent (overall RSDs of 10–37 percent). In summary, although the differences between matrices were generally less than 20 percent, statistical analysis shows that the highest overall mean recovery for the analytes was in the secondary wastewater effluent matrix, followed by the primary wastewater effluent matrix, and the reagent water, with the lowest overall recovery in the surface-water matrix.

Although summaries using parametric statistics are the primary comparative performance descriptors described in this report, also shown in table 16 are two nonparametric statistical descriptors: (1) median recoveries, which compare well with the mean recoveries except in the secondary wastewater effluent matrix for the analytes and in the surface-water matrix for the IDSs; and (2) relative F-pseudosigma (RF$_\sigma$), a "robust" indicator of relative variation based on the interquartile range of the data about the median (Hoaglin and others, 1983; see definition at *http://bqs.usgs.gov/ibsp/regress.shtml*, accessed April 2012). Unlike the RSD, the RF$_\sigma$ is not strongly influenced by extreme outliers in the data distribution. If the distribution is Gaussian, or nearly so, then RSD and RF$_\sigma$ are expected to be similar in magnitude because the variation will be (nearly) symmetric about the mean and median, which themselves should be (nearly) identical. Such is the case for the overall IDS data from the reagent-water or primary wastewater effluent matrices. The overall RF$_\sigma$ values are <23 percent for the analytes and <30 percent for the IDSs in these four matrices, and are substantially less than the corresponding overall RSDs that were strongly influenced by a few unusually low or high recoveries in several of the matrices.

Long-Term Estimates of Method Performance

Although the method-validation tests described previously in the "Primary Validation Matrices" section provided a consistent and comparable treatment using four different sample matrices, longer-term assessment of method performance is available from quality-assurance and

Table 15. Statistical comparison of recoveries between validation matrices and between fortification levels for method analytes and isotope-dilution standard compounds. The Wilcoxon rank-sum test was used to determine whether the distributions being compared came from the same population. For each comparison, the matrix or fortification level with significantly higher recovery is named (p-value less than 0.05).

[E, secondary wastewater effluent matrix; H, high-fortification level; L, low-fortification level; P, primary wastewater effluent matrix; R, reagent-water matrix; S, surface-water matrix; X, not reported because an ambient concentration exceeded 300 percent of the fortified amount, which produced substantially skewed recoveries in one or both matrices; --, no statistical difference between the distributions. Some results might have additional bias in statistical comparisons due to concentrations in the ambient sample from 25 to 150 percent (**bold** values) or from 151 to 300 percent (***bold italicized*** values) of the fortified amount]

Compound	R·S·L	R·S·H	R·E·L	R·E·L	R·E·H	R·E·H	R·S·L	R·S·H	S·E·L	S·E·L	S·E·H	S·E·H	E·P·L	E·P·L	E·P·H	E·P·H	R·Both	S·Both	E/P·Both
Method analytes																			
11-Ketotestosterone	R	—	R	R	E	E	R	R	P	P	P	P	—	E	E	E	—	L	***L***
17-*alpha*-Estradiol	E	—	E	E	E	E	P	P	—	—	P	P	—	H	H	H	—	—	L
17-*alpha*-Ethynylestradiol	R	R	R	R	R	R	—	E	E	E	E	E	E	—	E	E	—	—	H
17-*beta*-Estradiol	—	—	—	—	E	E	R	R	E	E	E	E	E	X	X	X	H	—	X
3-*beta*-Coprostanol	—	S	E	S	E	S	X	X	X	X	X	X	X	X	X	X	X	H	X
Androstenedione	S	S	E	E	E	S	X	X	X	X	X	X	X	X	X	X	L	H	X
Bisphenol A	R	R	E	E	E	P	X	X	E	E	E	P	E	X	E	X	H	***L***	X
Cholesterol	***S***	—	E	—	E	E	***S***	X	X	X	X	X	X	X	X	X	H	H	X
cis-Androsterone	S	R	—	R	E	E	S	S	S	S	S	S	X	X	X	X	L	H	X
Dihydrotestosterone	—	—	E	—	E	E	S	S	***S***	*S*	*S*	*S*	*E*	*R*	*E*	E	—	—	X
Epitestosterone	S	S	S	*P*	S	S	S	S	S	S	S	—	S	P	E	—	L	—	H
Equilenin	—	R	R	—	E	R	S	S	—	E	E	—	E	P	P	P	H	H	—
Equilin	S	S	E	—	E	E	E	E	—	E	—	E	E	*E*	P	*E*	H	L	—
Estriol	R	R	S	—	S	—	—	E	—	E	—	E	*P*	—	*P*	*P*	H	—	—
Estrone	***R***	—	***R***	—	***R***	—	E	E	—	E	E	*P*	*P*	—	*P*	*P*	—	—	—
Mestranol	—	—	E	R	E	E	E	E	—	E	—	E	E	—	E	E	H	—	H
Norethindrone	R	R	R	—	R	—	—	E	—	E	E	—	—	—	E	—	H	—	—
Progesterone	R	R	X	X	R	R	E	E	—	E	X	X	X	P	X	E	—	H	X
Testosterone	—	—	X	X	—	—	S	S	—	S	X	X	X	S	X	—	—	H	X
trans-Diethylstilbestrol	R	—	R	R	R	R	E	E	E	E	E	E	E	E	E	E	—	H	H
Isotope-dilution standards																			
16-Epiestriol-*d₂*	S	—	R	R	R	P	S	S	S	—	S	S	—	—	—	P	H	—	—
17-*alpha*-Ethynylestradiol-*d₄*	—	R	R	R	R	R	—	R	—	—	S	—	E	E	E	E	H	L	L
17-*beta*-Estradiol-¹³C₆	R	R	R	R	R	R	E	R	E	—	S	S	E	E	E	E	H	—	L
Bisphenol A-*d₁₆*	R	R	R	R	R	R	S	R	—	S	P	—	P	P	P	P	H	—	H
Cholesterol-*d₇*	R	—	R	S	R	R	S	—	S	S	S	S	E	—	E	E	L	H	—
Diethylstilbestrol-*d₈*	R	R	R	R	R	P	R	E	—	—	S	S	P	P	P	P	H	H	L
Estrone-¹³C₆	R	—	R	R	R	R	R	—	E	—	S	S	S	E	E	E	H	H	—
Medroxyprogesterone-*d₃*	R	E	R	R	R	—	E	E	E	P	P	P	P	E	E	E	H	—	L
Mestranol-*d₄*	R	R	R	R	R	R	S	S	S	—	S	S	—	E	S	S	H	—	L
Nandrolone-*d₃*	R	E	R	R	R	R	P	P	P	P	P	P	E	E	E	E	H	L	—

Table 16. Overall mean, median, relative standard deviation, and relative F-pseudosigma of recoveries of the method analytes and isotope-dilution standards for each validation matrix.

[All values in percent]

Overall recovery statistic	Reagent water	Surface water	Secondary wastewater effluent	Primary wastewater effluent	All four matrices
Method analytes					
Mean	97.9	93.0	110	99.5	100
Median	97.5	93.8	100	96.5	97.3
Relative standard deviation	9.8	29.3	36.7	25.0	28.5
Relative F-pseudosigma	6.5	22.1	12.6	15.5	12.4
Number of values	360	304[a]	340	184[a]	1,188[a]
Isotope-dilution standards					
Mean	81.7	61.3	73.9	67.0	71.4
Median	81.4	68.7	72.8	64.2	74.1
Relative standard deviation	9.6	38.1	20.5	22.1	24.9
Relative F-pseudosigma	9.2	29.3	14.3	23.4	19.5
Number of values	180	160	170	160	670

[a]Analytes having an ambient concentration that exceeded 300 percent of the amount spiked were excluded from the calculations. The matrices are described in the "Primary Validation Matrices" section.

quality-control data associated with numerous field samples prepared and analyzed during custom application as research methods to support a variety of USGS field projects. Data for six analytes and five IDSs presented in this section are associated with 247 filtered-water samples analyzed by LS 2434 and 578 unfiltered-water samples analyzed by LS 4434 during 2009–2010. Data for the remaining method analytes and IDS compounds are from samples (92 samples by LS 2434 and 316 samples by LS 4434) prepared in 2010 only, following the implementation of substitute IDS compounds for reasons described in section 10.7. The performance data are provided from laboratory reagent-water spike (LRS) and laboratory reagent-water blank (LRB) samples analyzed with each set of 10 (or more) environmental samples (see section 13). In addition, IDS recoveries provide a measure of method performance (comparable to surrogate recoveries) in the wide variety of sample matrices submitted for analysis that ranged from source groundwater to WWTP influent and animal feeding operation matrices. All samples were fortified with approximately 100 ng/L of nine IDS compounds and 10,000 ng/L of cholesterol-d_7, assuming a sample volume of 0.5 L. Sample volume-summary information also is included in some of the performance data tables to show the range of sample volumes processed.

Analyte Method Recoveries in Laboratory Reagent-Water Spike Samples

In addition to the fortified reagent-water replicates analyzed as a specific method validation matrix ("Primary Validation Matrices" section), as many as 113 laboratory reagent-water spike samples were analyzed in conjunction with custom sample analyses; these samples provide an

estimate of method performance in this reagent matrix over an extended time period (table 17). All LRSs were fortified with 25 ng/L of 17 method analytes, 250 ng/L of bisphenol A, 800 or 2,500 ng/L of 3β-coprostanol, and 2,500 ng/L of cholesterol, assuming a 0.5-L sample volume. These fortification concentrations fall between those used for the low-level (10 ng/L for most analytes) and high-level (100 ng/L) reagent-water validation replicates.

Analyte method recoveries in the LRSs relative to IDS absolute recoveries are shown in figure 9; these plots also include recoveries for the reagent-water validation replicates shown in figure 8. Most analyte recoveries fell within 60–120 percent, with only equilin, estriol, and 11-ketotestesterone having more than one recovery less than 60 percent. Sixteen of the analytes had more than one recovery greater than 120 percent, but only progesterone, dihydrotestosterone, epitestosterone, and *cis*-androsterone had more than one recovery that exceeded 150 percent, and these occurred when the recovery for the corresponding non-exact IDS analog was in the low range for this matrix.

Mean analyte recoveries in the LRSs ranged from 84 to 104 percent (table 17) and generally are similar to those obtained from the low and high level reagent water validation replicates (tables 12 and 13). The RSDs were ≤25 percent for all analytes in the LRSs except estriol (28 percent), 11-ketotestosterone (29 percent), and progesterone (36 percent), and, not unexpectedly, were greater than the RSDs from the reagent-water validation tests. The RF$_σ$ values were <16 percent for all analytes in the LRSs and are substantially less than the corresponding RSDs for estriol and 11-ketotestesterone that were strongly influenced by unusually low recoveries in as many as five LRS samples (fig. 9).

The reason for the low estriol and 11-ketotestesterone recoveries in several LRS samples (and four matrix-spike

Table 17.   Statistical summary of analyte and isotope-dilution standard compound recoveries from 113 laboratory reagent-water spike (LRS) samples prepared in 2009–2010 or from 51 LRS samples prepared in 2010 for compounds quantified with new isotope dilution standards.

[N, number of samples; RSD, relative standard deviation; Fortification level was 25 nanograms per liter (ng/L) for 17 analytes, 250 ng/L for bisphenol A, 800 or 2,500 ng/L for 3-*beta*-coprostanol, and 2,500 ng/L for cholesterol assuming a 0.5-L sample volume. Isotope-dilution standards were fortified at 100 ng/L, except cholesterol-d_7, which was fortified at 10,000 ng/L]

Compound	N	Mean recovery (percent)	RSD (percent)	Median recovery (percent)	Relative F-pseudosigma (percent)	Minimum recovery (percent)	Maximum recovery (percent)
			Method analytes				
11-Ketotestosterone	51	85.7	29.1	90.7	8.3	8.1	111
17-*alpha*-Estradiol	51	101	12.9	99.2	5.4	81.8	149
17-*alpha*-Ethynylestradiol	113	95.2	7.1	94.7	6.7	76.8	124
17-*beta*-Estradiol	51	100	13.0	97.7	5.0	86.3	148
3-*beta*-Coprostanol	112[a]	102	14.3	98.9	15.7	64.8	152
4-Androstene-3,17-dione	51	93.7	8.0	93.3	8.0	78.2	115
Bisphenol A	113	91.7	12.9	88.3	13.2	72.3	128
Cholesterol	112[a]	94.8	10.1	93.7	7.0	74.1	134
cis-Androsterone	51	104	21.0	98.4	11.1	81.2	178
Dihydrotestosterone	51	97.6	20.5	92.4	10.9	76.8	170
Epitestosterone	51	101	16.6	95.7	8.0	83.7	163
Equilenin	51	87.3	15.2	85.3	8.1	60.1	138
Equilin	51	83.7	17.4	82.6	12.4	50.6	124
Estriol	51	83.9	28.4	89.2	12.6	11.4	117
Estrone	51	100	5.6	101	5.4	90.2	121
Mestranol	113	99.5	7.5	98.7	4.9	85.2	132
Norethindrone	51	96.6	7.7	96.4	7.5	84.2	118
Progesterone	51	90.1	36.1	83.5	14.6	58.7	211
Testosterone	51	101	14.5	96.7	8.2	83.1	151
trans-Diethylstilbestrol	113	90.3	6.6	89.9	6.8	68.0	105
			Isotope-dilution standards				
16-Epiestriol-d_2	51	64.5	27.0	67.4	12.0	10.0	93.4
17-alpha-Ethynylestradiol-d_4	113	78.9	10.4	79.0	10.0	53.7	101
17-*beta*-Estradiol-$^{13}C_6$	51	79.2	14.0	79.9	9.7	51.1	110
Bisphenol A-d_{16}	113	81.4	13.4	82.2	10.0	36.5	115
Cholesterol-d_7	113	68.1	13.2	68.0	13.5	46.6	97.7
Diethylstilbesterol-d_8	113	66.0	12.5	64.3	13.5	50.3	91.3
Estrone-$^{13}C_6$	51	77.0	12.8	76.0	12.2	58.8	115
Medroxyprogesterone-d_3	51	68.5	20.9	70.9	14.3	28.2	92.6
Mestranol-d_4	113	74.2	9.8	73.5	9.5	53.8	97.4
Nandrolone-d_3	51	73.3	15.3	72.7	9.1	43.3	96.3
		(milliliters)	(percent)	(milliliters)	(percent)	(milliliters)	(milliliters)
Sample volume	113	460	2.2	463	0.7	425	473

[a]One high-biased value omitted because of analyte injection carryover from preceeding analysis of an influent sample.

samples; see "Compound Recoveries in Other Spiked-Matrix Samples" section) is unknown, but is suspected to be attributed to derivatization limitations. Most analytes and IDS compounds contain one or (more commonly) two C–OH or C=O functional groups (figs. 1 and 2) that are converted to trimethylsilyl derivatives upon reaction with *N*-methyl-*N*-trimethylsilyl trifluoroacetamide (MSTFA) (fig. 6), whereas estriol and 16-epiestriol-d_2 have three C–OH groups and 11-ketotestosterone has two C–OH groups and one C=O group that require conversion to trimethylsilyl derivatives. Thus, it is possible that derivatization of all three functional groups on estriol and 11-ketotestosterone was incomplete in these particular samples compared to their corresponding IDS compounds (16-epiestriol-d_2 and nandrolone-d_3, respectively) and compared to the other analytes and their corresponding

IDS compounds. If derivatization of the analyte is complete in the calibration standards but not in a given sample matrix, whereas derivatization of the corresponding IDS is complete in both the calibration standards and the given sample matrix, then the analyte's performance is not being well emulated by its IDS in that sample matrix and the determined analyte concentration (or recovery) in that sample will be biased low.

Estriol was determined to elute incompletely (15 percent maximum estriol retention) from a larger 2-g Florisil cleanup column used in a complementary method developed at the NWQL for the determination of steroid hormones in solids (method description summarized in Lee and others, 2011) when using 25 mL of 5-percent methanol in dichloromethane solution. Although it can not be completely discounted, the extensive loss of estriol and 11-ketotestosterone in the few

LRS (and matrix spike) samples probably did not occur when using the smaller 1-g Florisil SPE cleanup columns in this water method. Improvements in matrix-specific method recoveries for estriol are expected by the recent substitution of the exact isotopic analog estriol-d_4 for 16-epiestriol-d_2 (see section 10.7).

Isotope-Dilution Standard Performance

The IDS compounds have a dual role in this method; primarily they are used to automatically adjust quantitation (concentration) of method analytes in relation to sample-specific absolute recovery. In addition, the IDS recoveries provide a direct measure of sample-specific absolute (not IDS-corrected) recovery for those analytes that have exact-analog IDS compounds. Likewise, sample-specific recoveries for those IDSs that are structurally similar, but not exact, isotopic analogs provide an estimate of absolute recovery for the corresponding determined analyte (see table 7). Thus, IDS recovery data provide useful sample-specific information for the method relevant to efficiency of the extraction, evaporation, cleanup, derivatization, and analysis procedures. Accumulation of IDS recovery data, thus, provides an indicator of overall method performance as absolute recovery.

Isotope-Dilution Standard Absolute Recoveries in Laboratory Reagent-Water Spike Samples

Mean IDS recoveries in the LRS samples range from 64 to 81 percent (table 17), and are similar overall with those obtained for the reagent-water validation matrix (especially the low-level spikes) shown in tables 12 and 13. Mean recoveries were somewhat lower in the LRSs for 16-epiestriol-d_2, medroxyprogesterone-d_3, and nandrolone-d_3 compared to the reagent-water validation matrix. The RSDs in the LRS samples were ≤21 percent except for 16-epiestriol-d_2 (27 percent), and greater than observed for the reagent-water validation.

Isotope-Dilution Standard Absolute Recoveries in Laboratory Reagent-Water Blank and Field-Blank Samples

Table 18 summarizes the IDS recoveries in as many as 115 laboratory reagent-water blank (LRB) samples and compares them with those obtained from as many as 15 field-blank samples prepared for analysis by LS 2434 (field filtered) and as many as 70 field blank samples prepared for analysis by LS 4434 (no field filtration). The LRB samples were prepared using reagent water from the Solution 2000 water-purification system as used for the reagent-water replicates and LRS samples. The sample-specific reagent water used for submitted field blanks was not identified to the NWQL, but was known to come from at least the following sources: organic blank water items N1580 or N1590 from the USGS National

Field Supply Service, or prepared by Water Science Center personnel using an in-house water-purification system.

As expected, mean IDS recoveries and RSDs for the LRB samples (table 18) are similar to those for the LRS samples (table 17). Mean IDS recoveries for the LRB and field-blank samples also are similar, except for the apparent lower mean recoveries of nandrolone-d_3 and especially of 16-epiestriol-d_2 in the field-blank samples for LS 2434. However, the small number of field-blank samples submitted for LS 2434 might not be representative for several IDS compounds. The IDS compounds are fortified to the sample just before extraction at the NWQL, so the field-filtration process itself is not the direct cause for the lower recoveries for these two IDSs in LS 2434 samples.

Isotope-Dilution Standard Absolute Recoveries in Field-Sample Matrices

A summary of IDS recoveries in ambient field-sample matrices that were analyzed by using LS 2434 and LS 4434 is shown in table 19. Mean IDS recoveries for field samples ranged from 48 to 85 percent and were similar between LS 2434 and LS 4434 samples for most IDS compounds. A distinct exception was medroxyprogesterone d_3, which had substantially lower recoveries and higher variation in LS 4434 field samples compared to LS 2434 field samples. Nandrolone-d_3 also had somewhat lower recoveries and greater variability in LS 4434 samples compared to LS 2434 samples. For all IDS compounds except 16-epiestriol-d_2, RSDs were greater for LS 4434 field samples than for LS 2434 field samples. All RF_σ values for LS 4434 field samples were at least slightly greater than those for LS 2434 field samples; with the greatest difference for medroxyprogesterone-d_3. Mean IDS recoveries in field samples generally were similar to those observed for LRB samples (table 18) and LRS samples (table 17), although variability in the field-sample matrices was greater than observed with the reagent-water matrices (LRB and LRS samples) for some IDSs.

Table 20 summarizes IDS recoveries grouped by water matrix (medium) type and laboratory schedule. For most samples, the water matrix classification listed is based on the NWIS medium code (shown in table 20) used by field staff on the Analytical Services Request form (see medium code definitions in U.S. Geological Survey, 2011a). However, three matrix types suspected to be "more complex" were grouped separately from the user-assigned NWIS medium code based on the station name or other sample-source information provided on the Analytical Services Request form because NWIS medium codes are not available that uniquely identify these three matrix types. These matrix types are (1) WWTP influent (includes primary effluent) samples, (2) hog manure slurry samples, and (3) surface-water samples believed to have been affected by a hog manure-waste spill event. Matrix types listed in table 20 are not represented for all 10 IDS compounds because samples for some matrix types were prepared before

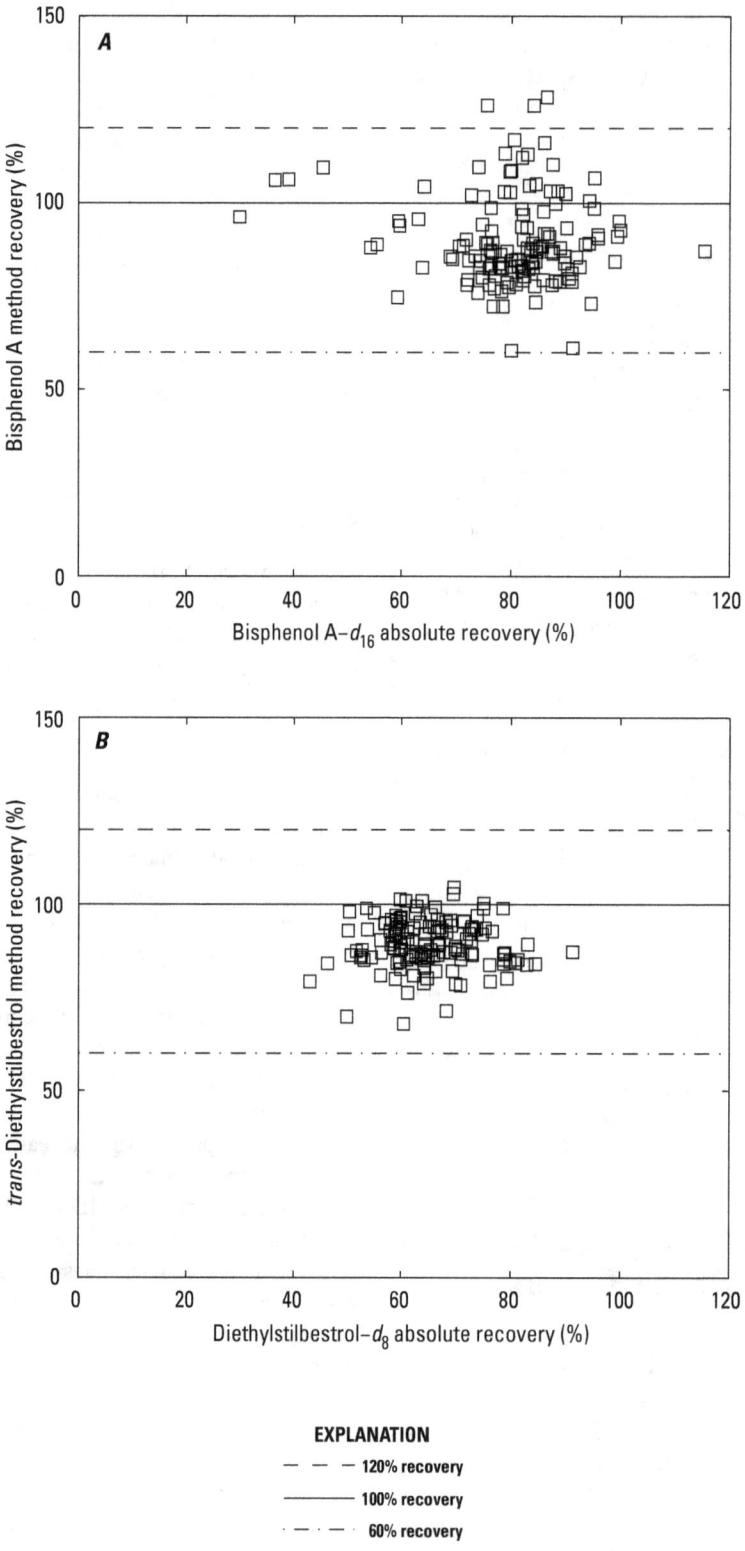

Figure 9. Relation between analyte method recoveries and isotope-dilution standard absolute recoveries in percent (%) in laboratory reagent-water spike samples prepared in 2009–2010 for six analytes or in 2010 only for 14 analytes.

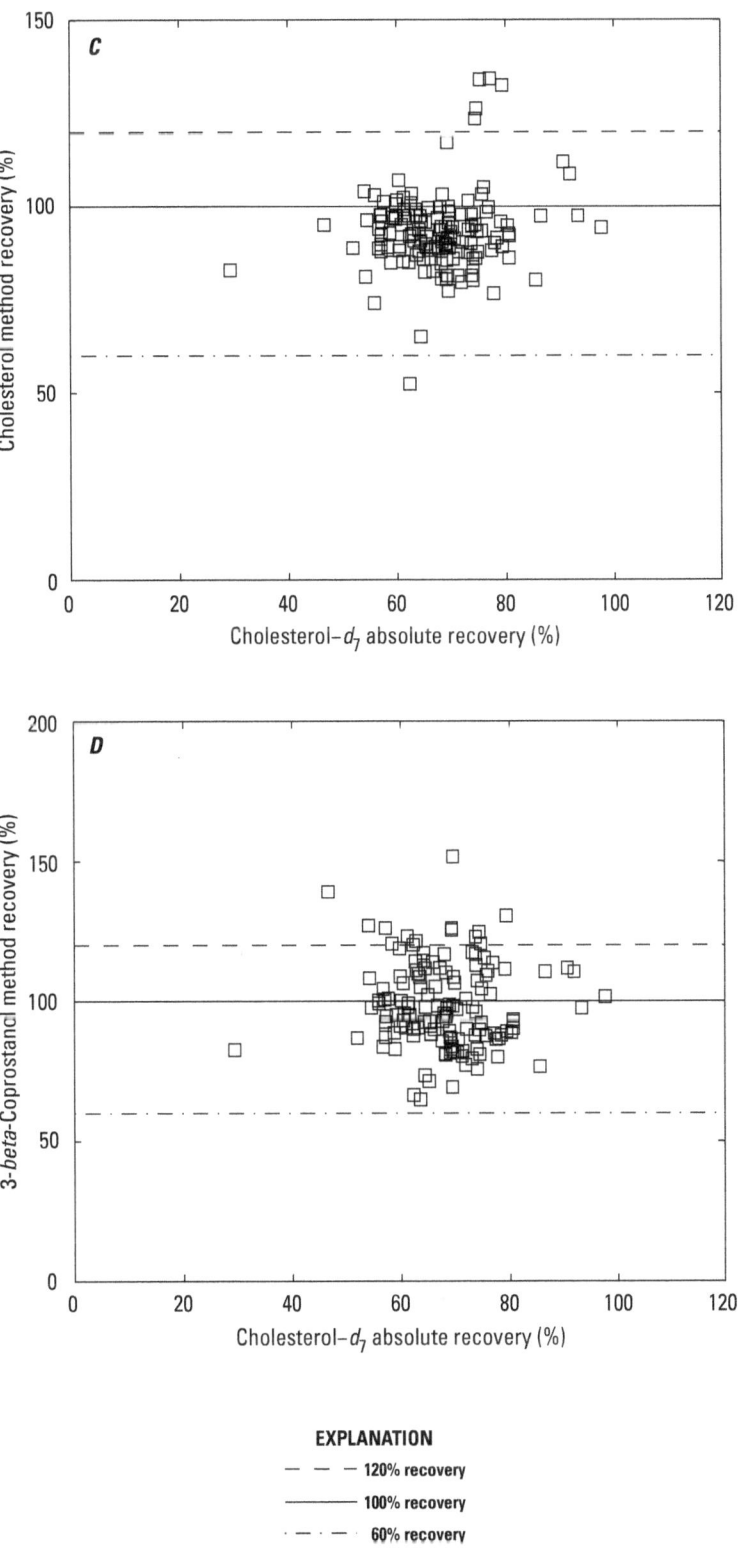

Figure 9. Relation between analyte method recoveries and isotope-dilution standard absolute recoveries in percent (%) in laboratory reagent-water spike samples prepared in 2009–2010 for six analytes or in 2010 only for 14 analytes.— Continued

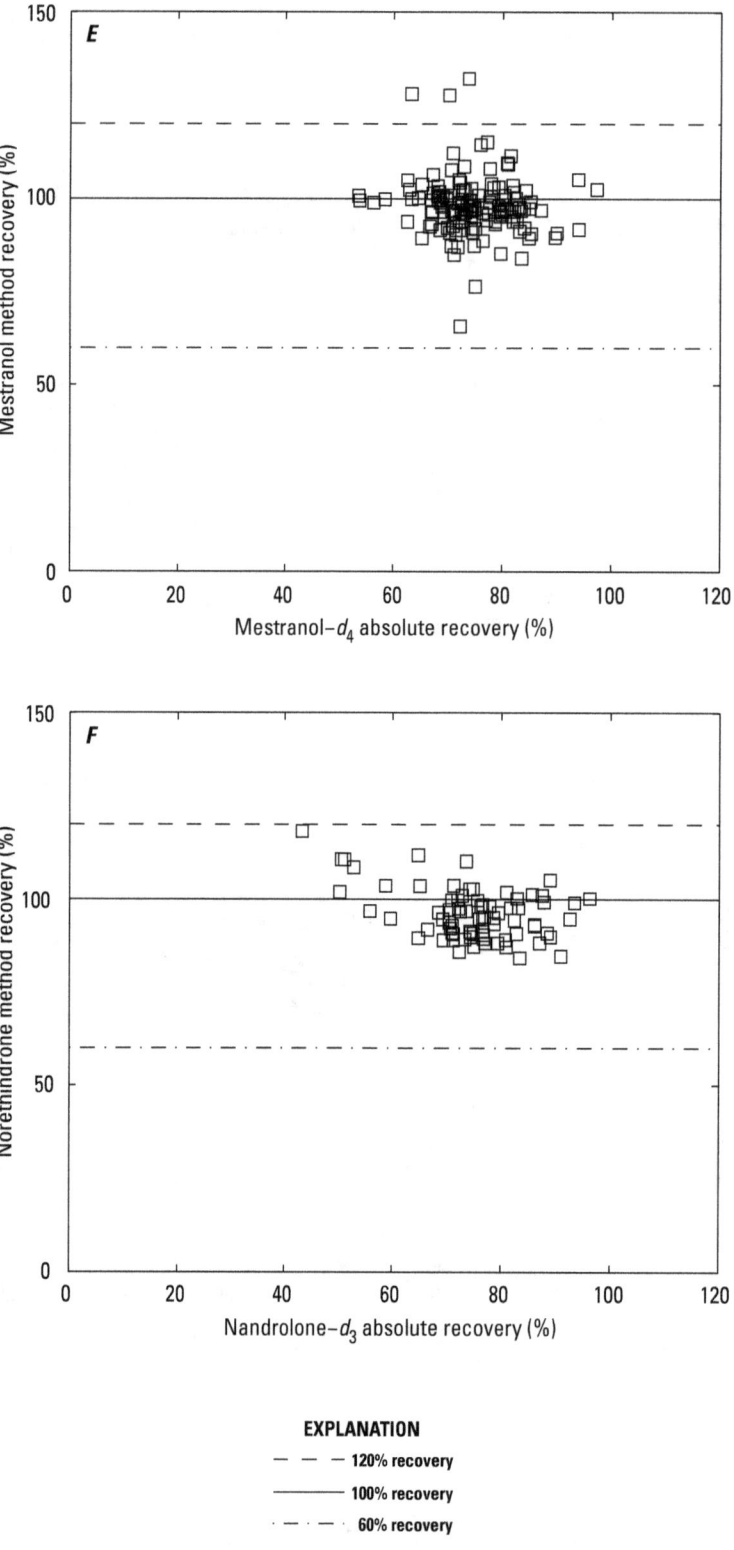

EXPLANATION

— — — 120% recovery

———— 100% recovery

— · — · 60% recovery

Figure 9.   Relation between analyte method recoveries and isotope-dilution standard absolute recoveries in percent (%) in laboratory reagent-water spike samples prepared in 2009–2010 for six analytes or in 2010 only for 14 analytes.— Continued

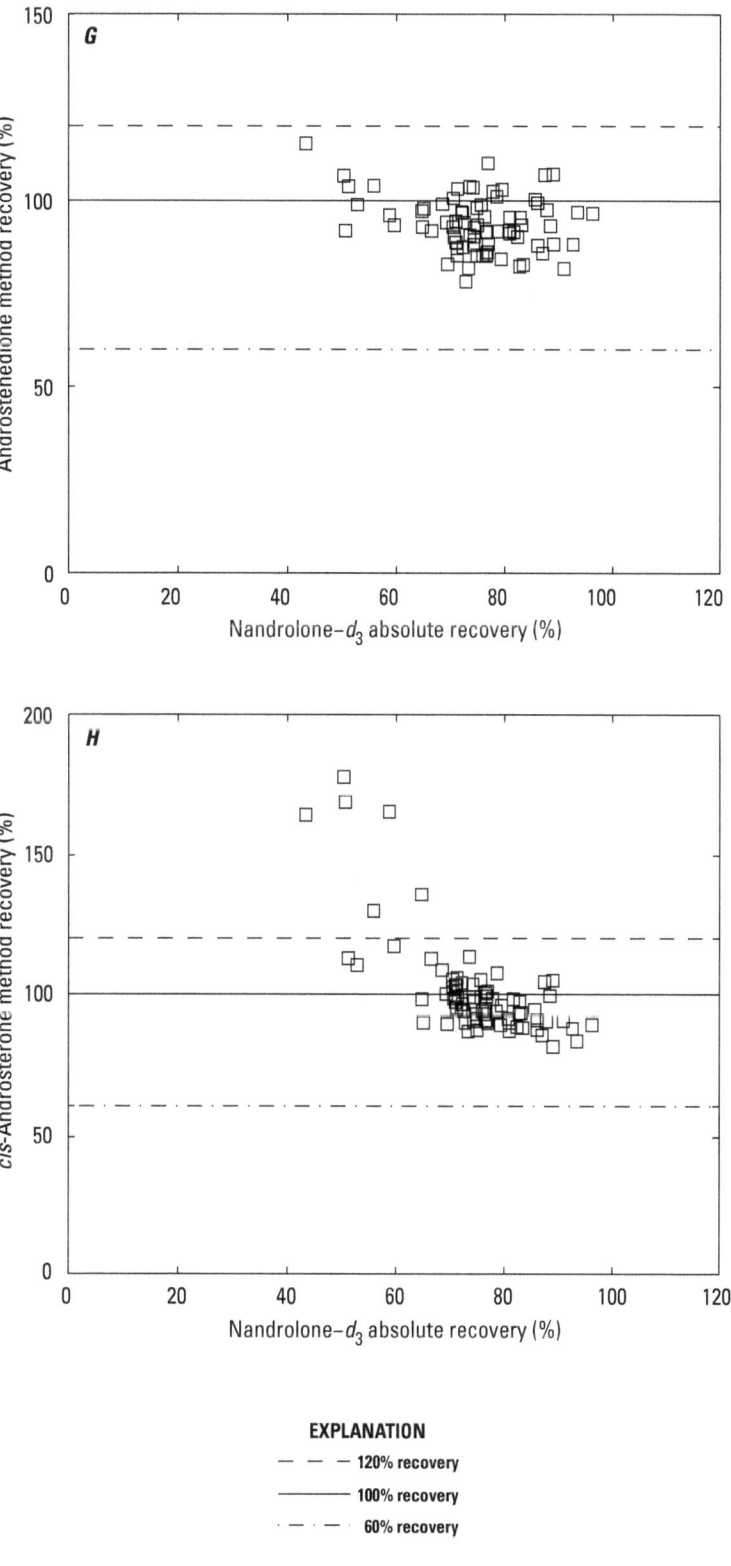

Figure 9. Relation between analyte method recoveries and isotope-dilution standard absolute recoveries in percent (%) in laboratory reagent-water spike samples prepared in 2009–2010 for six analytes or in 2010 only for 14 analytes.— Continued

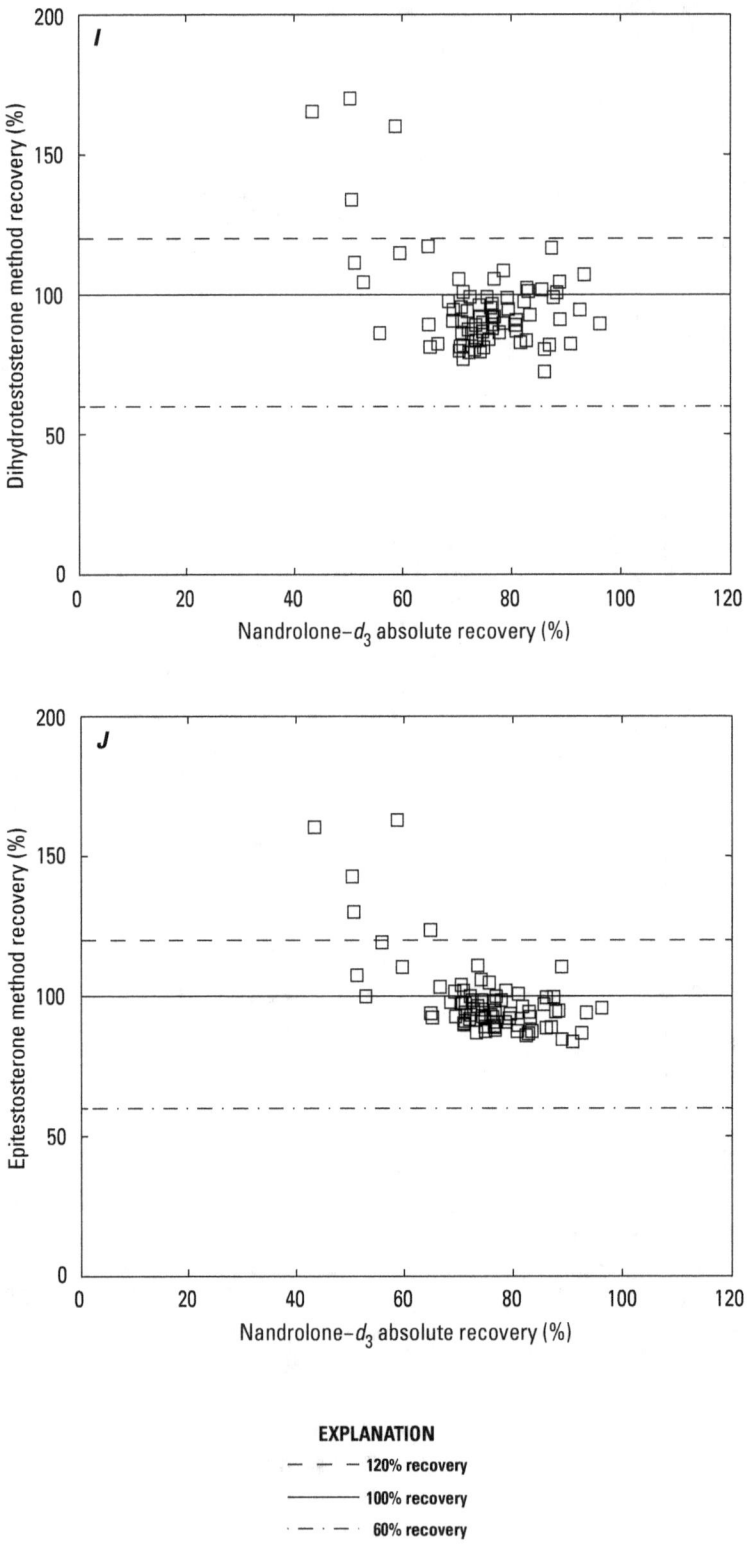

Figure 9. Relation between analyte method recoveries and isotope-dilution standard absolute recoveries in percent (%) in laboratory reagent-water spike samples prepared in 2009–2010 for six analytes or in 2010 only for 14 analytes.— Continued

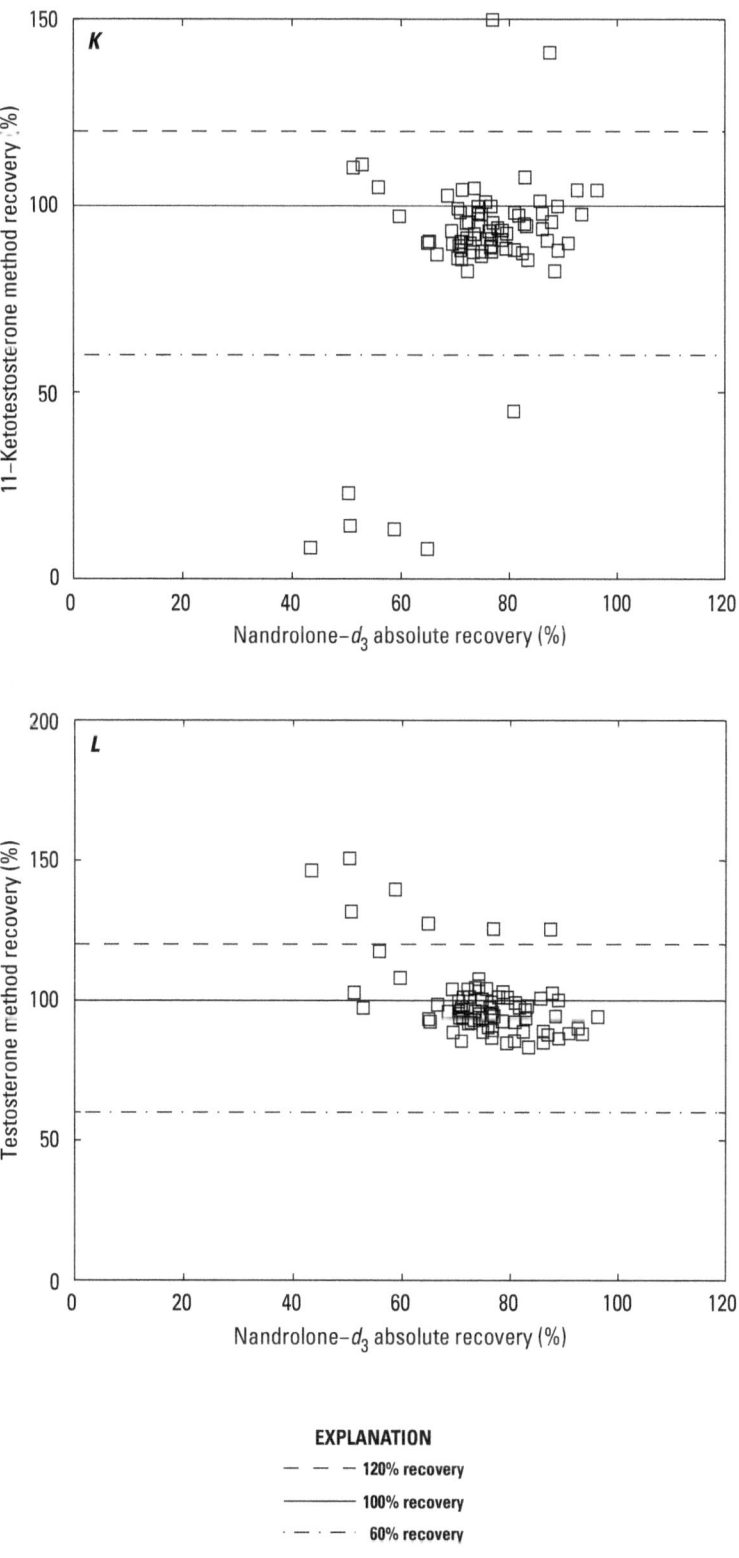

Figure 9.   Relation between analyte method recoveries and isotope-dilution standard absolute recoveries in percent (%) in laboratory reagent-water spike samples prepared in 2009–2010 for six analytes or in 2010 only for 14 analytes.— Continued

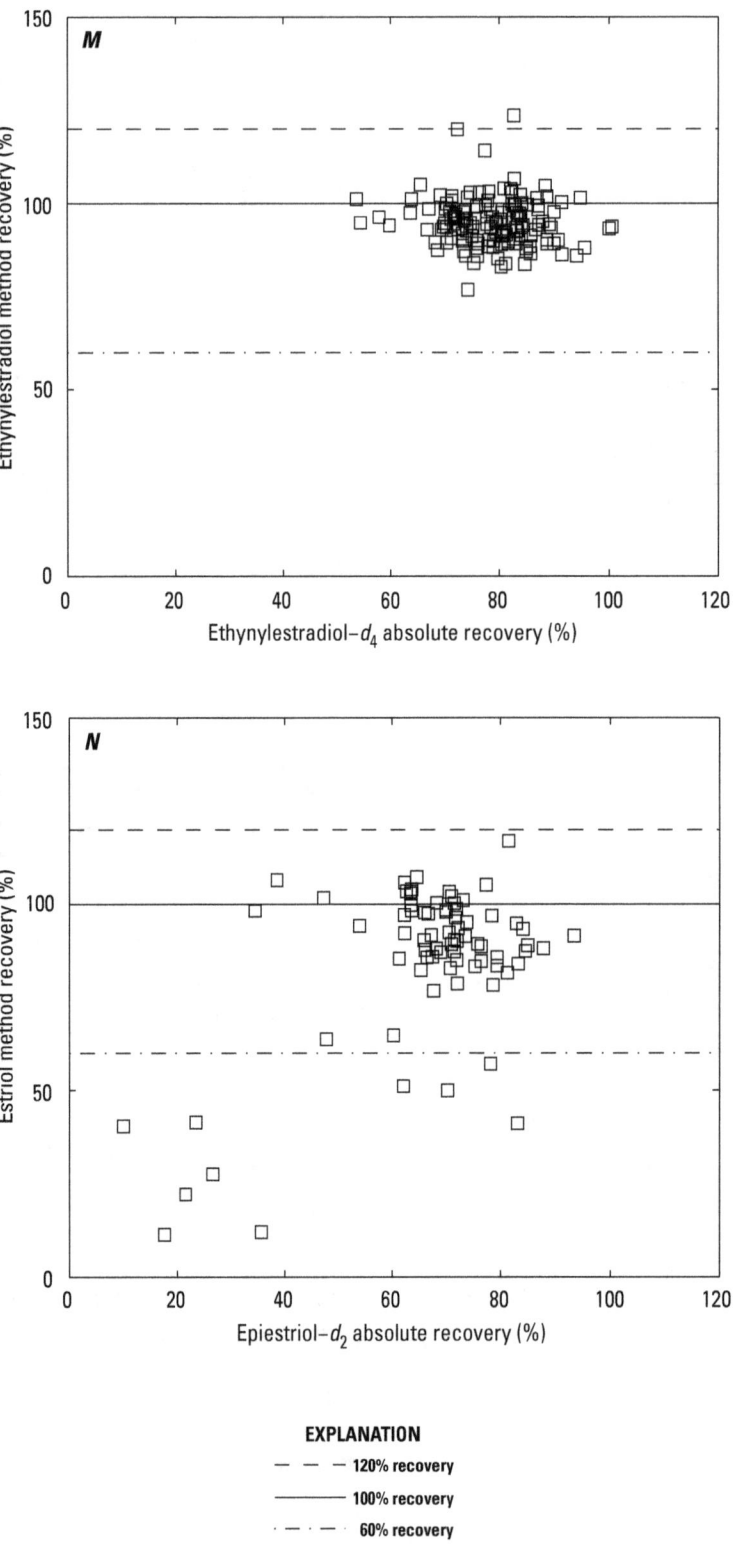

Figure 9. Relation between analyte method recoveries and isotope-dilution standard absolute recoveries in percent (%) in laboratory reagent-water spike samples prepared in 2009–2010 for six analytes or in 2010 only for 14 analytes.—Continued

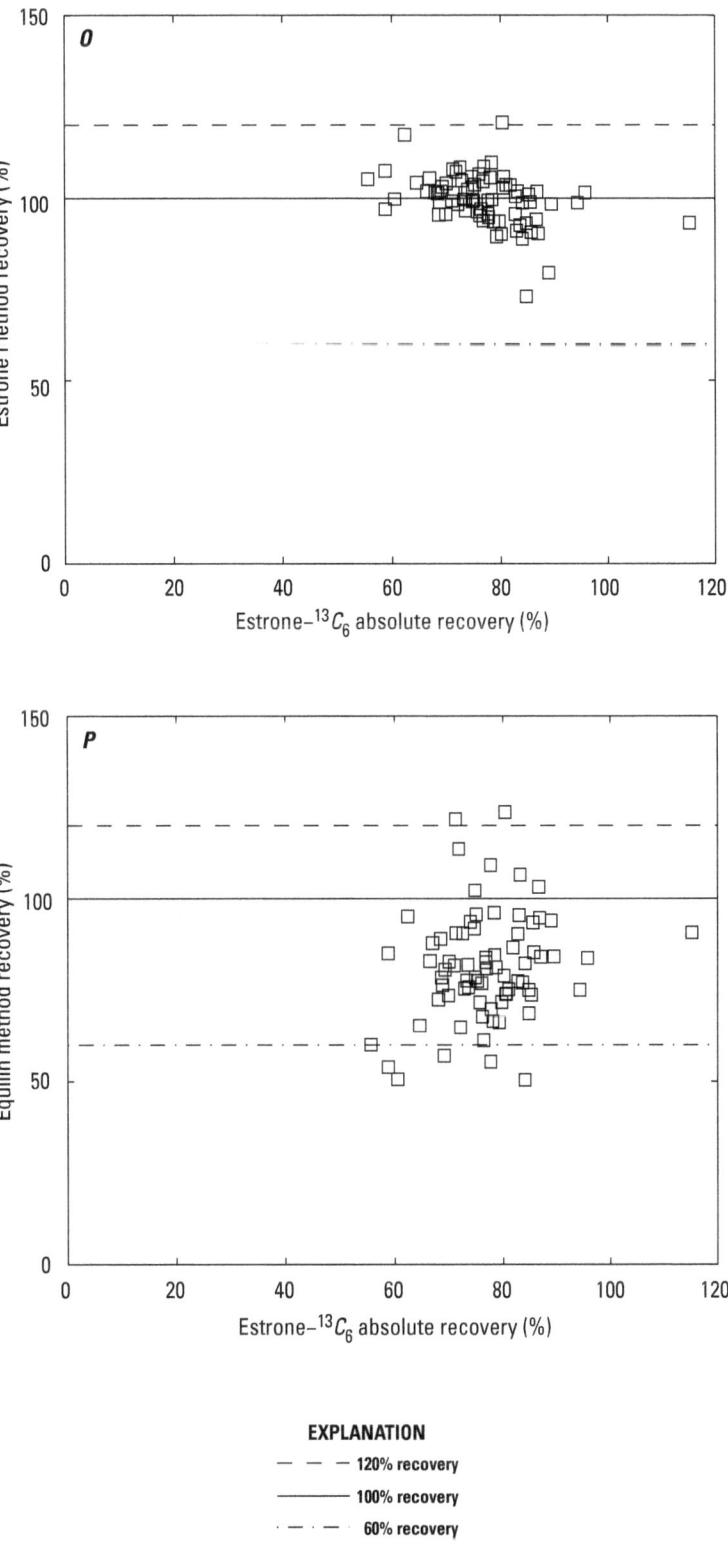

Figure 9.   Relation between analyte method recoveries and isotope-dilution standard absolute recoveries in percent (%) in laboratory reagent-water spike samples prepared in 2009–2010 for six analytes or in 2010 only for 14 analytes.— Continued

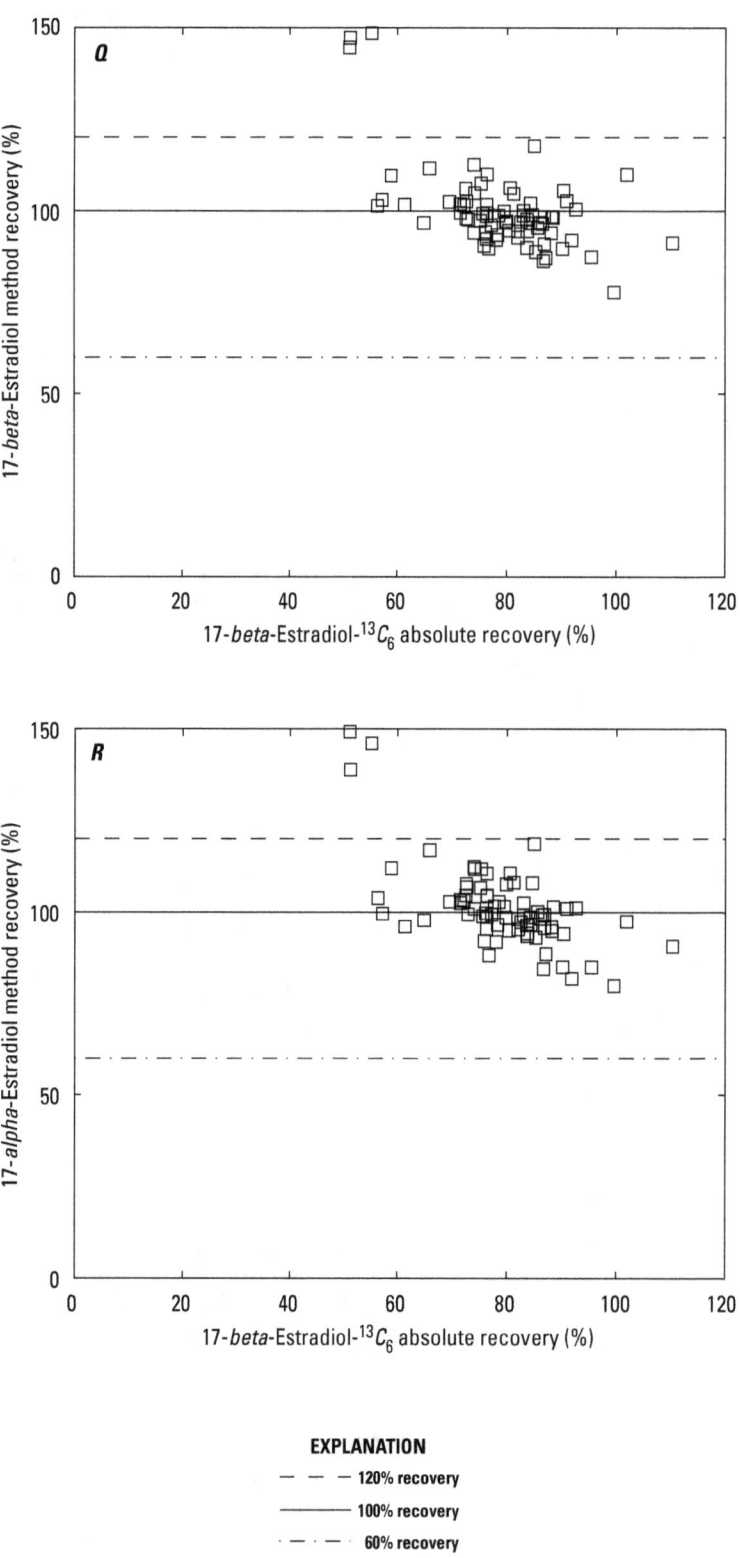

Figure 9. Relation between analyte method recoveries and isotope-dilution standard absolute recoveries in percent (%) in laboratory reagent-water spike samples prepared in 2009–2010 for six analytes or in 2010 only for 14 analytes.— Continued

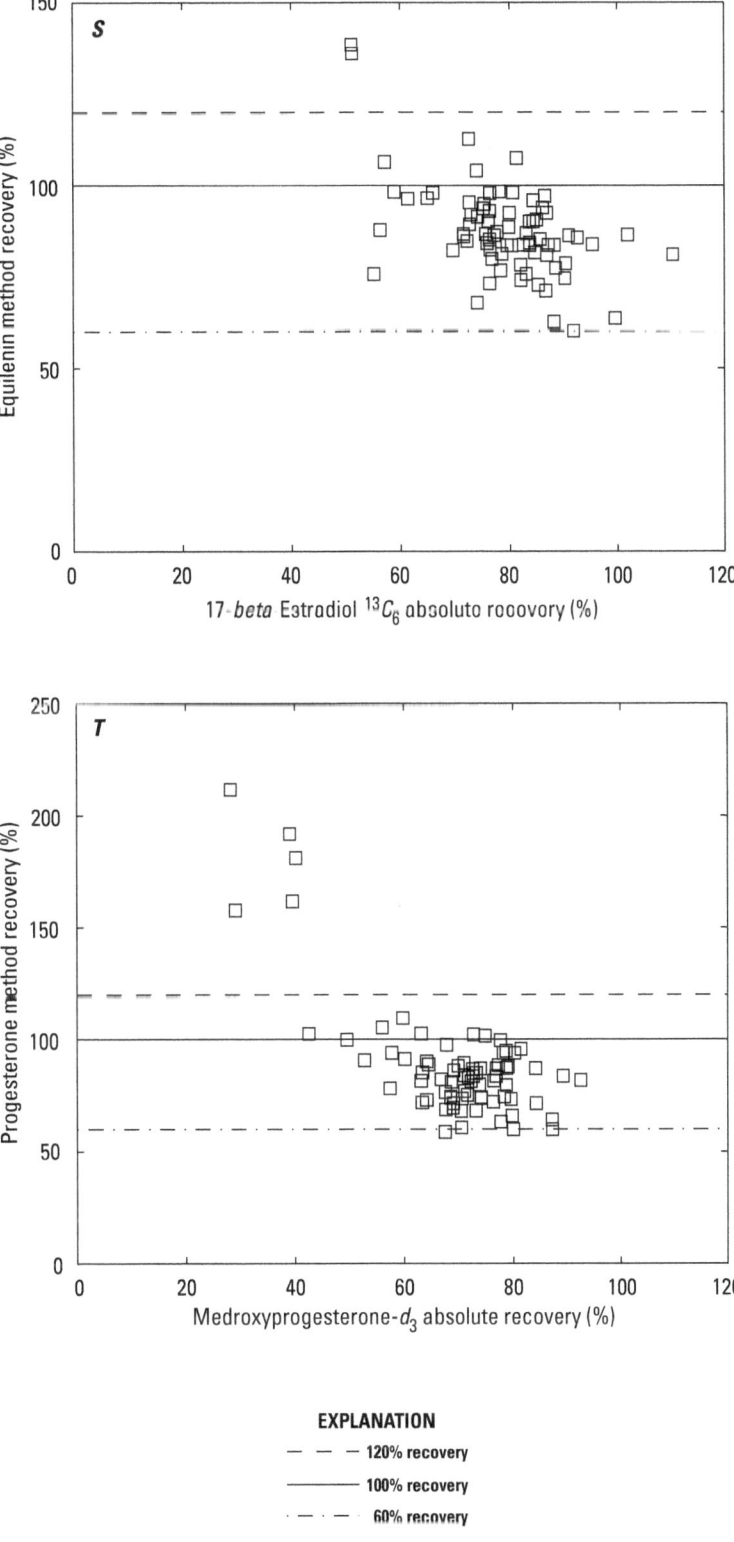

Figure 9. Relation between analyte method recoveries and isotope-dilution standard absolute recoveries in percent (%) in laboratory reagent-water spike samples prepared in 2009–2010 for six analytes or in 2010 only for 14 analytes.— Continued

implementation of substitute IDS compounds as described in section 10.7.

Mean IDS recoveries were greater than 60 percent (with many in the range of 70 to 90 percent) in most matrices and generally were of a similar magnitude between matrix types or between laboratory schedules within a matrix type. Similarly, RSDs were ≤25 percent for most matrices and of comparable magnitude for many matrix types regardless of laboratory schedule. These data indicate that the analytical method is applicable to diverse water matrix types. However, recoveries were substantially lower or more variable, or both, for one or more IDS compounds in some matrix types, especially those having a more "complex" makeup (higher amounts of suspended solids, DOC, and coextracted organic matter compared to other matrices), including WWTP influent (especially unfiltered samples), hog manure slurry, and sludge samples. In addition, 16-epiestriol-d_2 had somewhat lower

recoveries in the groundwater matrix for LS 2434, and medroxyprogesterone-d_3 had lower recoveries (with high variability) in the surface-water matrices for LS 4434.

The assumption in this IDQ-based method is that the absolute recovery of progesterone in a given sample matrix is closely emulated by the recovery of medroxyprogesterone-d_3. As noted previously in the "Surface Water" section for the surface-water validation matrix, recoveries of medroxyprogesterone-d_3 were particularly low in some surface-water matrices. Based on the validation and matrix-spike (see "Compound Recoveries in Other Spiked-Matrix Samples" section) test samples, the absolute medroxyprogesterone-d_3 recovery was similar to the absolute progesterone recovery in some matrices. When they differed, the progesterone absolute recovery typically was less than the medroxyprogesterone-d_3 recovery (sometimes substantially lower), leading to progesterone method recoveries that remained well below the 100 percent

Table 18. Statistical summary of recoveries for isotope-dilution standard compounds from laboratory reagent-water blank (LRB) and field-blank samples submitted for laboratory schedules 2434 (2434FB) and 4434 (4434FB) in 2009–2010 or 2010 only.

[N, number of samples; RSD, relative standard deviation; Isotope-dilution standards were fortified at 100 nanograms per liter (ng/L), except cholesterol-d_7, which was fortified at 10,000 ng/L]

Isotope-dilution standard	Sample type	N	Mean (percent)	RSD (percent)	Median (percent)	Minimum (percent)	Maximum (percent)
16-Epiestriol-d_2	LRB	52	63.2	27.7	65.0	12.4	91.9
	2434FB	4	33.4	63.3	24.9	19.4	64.4
	4434FB	22	73.0	16.2	71.3	51.8	99.7
17-*alpha*-Ethynylestradiol-d_4	LRB	115	78.0	10.9	77.0	58.1	99.5
	2434FB	15	81.9	16.6	78.1	68.2	127
	4434FB	70	82.3	16.4	82.0	54.1	143
17-*beta*-Estradiol-$^{13}C_6$	LRB	52	76.3	15.8	76.2	46.0	106
	2434FB	4	76.7	2.7	76.4	74.9	79.1
	4434FB	22	79.6	10.2	79.1	67.4	94.8
Bisphenol A-d_{16}	LRB	115	81.4	13.3	83.0	27.3	107
	2434FB	15	81.2	20.9	85.4	47.7	104
	4434FB	70	83.7	14.6	82.8	53.6	121
Cholesterol-d_7	LRB	115	67.6	14.1	67.2	17.4	96.8
	2434FB	15	69.9	9.7	70.0	59.0	87.2
	4434FB	70	70.9	13.4	70.1	50.6	102
Diethylstilbesterol-d_8	LRB	115	64.3	15.3	64.4	38.4	93.4
	2434FB	15	73.2	13.5	70.5	59.3	97.3
	4434FB	70	70.7	18.6	70.2	46.8	106
Estrone-$^{13}C_6$	LRB	52	74.2	12.2	74.3	51.9	91.8
	2434FB	4	73.3	4.0	73.8	69.3	76.2
	4434FB	22	80.4	12.7	80.5	64.0	97.4
Medroxyprogesterone-d_3	LRB	52	71.1	20.7	73.5	20.9	95.5
	2434FB	4	71.8	16.5	69.0	61.2	88.2
	4434FB	22	79.1	12.4	78.6	64.3	99.7
Mestranol-d_4	LRB	115	72.9	10.7	73.3	50.9	98.1
	2434FB	15	79.2	9.7	77.8	68.4	97.0
	4434FB	70	76.8	11.6	77.5	53.7	95.1
Nandrolone-d_3	LRB	52	73.1	15.0	74.2	45.8	96.3
	2434FB	4	63.1	16.2	62.5	52.0	75.4
	4434FB	22	81.4	10.6	83.3	63.7	96.0
			(milliliters)	**(percent)**	**(milliliters)**	**(milliliters)**	**(milliliters)**
Sample volume	LRB	115	459	2.5	463	408	475
	2434FB	15	451	7.3	462	373	512
	4434FB	70	606	27.1	657	296	948

optimum recovery even when using IDQ (see "Compound Recoveries in Other Spiked-Matrix Samples" section). Only occasionally was this condition reversed, leading to a high biased progesterone recovery. As a consequence, determined progesterone concentrations might be biased low in a given sample matrix, but are less likely to be biased high. For this and other reasons noted previously, progesterone concentrations are reported to NWIS as estimated.

The range of sample volumes (table 19) processed for the LS 4434 samples (43–1,025 mL) was substantially greater than the range processed for the LS 2434 samples (283–589 mL) because at least one matrix (a hog manure slurry) for LS 4434 clogged the GFF/C_{18} disk during extraction (43 mL extracted; the only matrix of those tested that was not amenable to complete sample-volume extraction) or because a 1-L HDPE bottle was used for several field projects. The extraction of sample volumes larger than 450 mL (to as much as 1 L) was not related to the occurrence of lower or more variable IDS recoveries; instead, matrix type affected the recoveries.

Cholesterol-d_7 Recoveries in Non-salted Reagent Water

Sodium chloride (salt) is added to all sample matrices in this method before extraction, and is done specifically to improve the absolute recoveries of cholesterol-d_7 and the two sterol analytes, cholesterol and 3β-coprostanol, in reagent-water matrices. In method testing and custom analysis applications before 2009, recoveries of cholesterol-d_7 were found to be especially poor (about 10 percent on average) in reagent-water-based matrices, including all field-blank samples and the spike samples submitted by the Organic Blind Sample Project of the USGS Branch of Quality Systems.

Loss of these three sterol compounds in unsalted reagent-water samples was determined to occur because of their incomplete isolation on the C_{18} disk during the solid-phase extraction step (section 9.7$_4$). This occurs because a substantial portion of the sterols passes through the C_{18} sorbent material (referred to as sorbent breakthrough) during SPE instead of being well retained by the sorbent. This finding was established by fortifying triplicate 0.5-L reagent-water samples contained in amber glass bottles with 12 IDSs (those used before the IDS changes implemented in 2010, see section 10.7) and 19 analytes (bisphenol A and its IDS were not in the method at that time). The samples were extracted using standard conditions. No sodium chloride was added to the water before extraction for this test.

The Florisil cleanup step (which was not causing the low sterol recoveries) was bypassed in this testing to limit the handling steps to the extraction, extraction reduction, and final derivatization steps. Analyte and IDS recoveries from five components relevant to the extraction process were determined to identify possible losses during this part

Table 19. Bias and variability of recoveries for isotope-dilution standard compounds from ambient field samples collected in 2009–2010 and analyzed by using laboratory schedule 2434 or 4434.

[N, number of samples]

Isotope-dilution standard	Laboratory schedule	N	Mean (percent)	RSD (percent)	Median (percent)	Relative F-pseudo-sigma (percent)	Minimum (percent)	Maximum (percent)
16-Epiestriol-d_2	2434	92	66.6	35.5	73.1	15.3	13.7	106
	4434	316	74.6	22.0	76.9	17.1	12.6	154
17-*alpha*-Ethynylestradiol-d_4	2434	247	82.1	14.4	80.8	13.5	56.5	115
	4434	578	79.7	19.3	79.1	14.9	39.0	148
17-*beta*-Estradiol-$^{13}C_6$	2434	92	83.7	10.9	81.6	9.1	68.6	105
	4434	316	73.8	19.0	76.8	17.7	2.4	118
Bisphenol A-d_{16}	2434	247	84.9	18.9	86.7	17.6	32.3	123
	4434	578	83.9	27.6	87.1	24.1	0.2	156
Cholesterol-d_7	2434	247	73.0	12.2	72.9	11.8	49.0	94.7
	4434	578	69.6	19.5	72.3	15.2	15.7	108
Diethylstilbesterol-d_0	2434	247	61.2	26.1	61.7	24.8	20.3	101
	4434	578	57.1	37.1	56.7	39.3	4.9	134
Estrone-$^{13}C_6$	2434	92	85.4	16.4	84.2	11.5	0.0	111
	4434	316	78.3	21.0	79.6	13.0	0.0	127
Medroxyprogesterone-d_3	2434	92	77.5	27.9	78.8	20.2	7.2	134
	4434	316	48.2	63.8	46.8	77.8	0.0	128
Mestranol-d_4	2434	247	79.9	11.4	80.8	10.8	56.1	104
	4434	578	77.8	17.3	77.9	12.7	30.6	166
Nandrolone-d_3	2434	92	81.5	22.0	83.3	15.1	1.6	125
	4434	316	67.5	34.0	67.4	27.6	0.6	170
			(milliliters)	(percent)	(milliliters)	(percent)	(milliliters)	(milliliters)
Sample volume	2434	247	438	7.1	440	6.4	283	589
	4434	578	590	31.0	587	33.5	43.1	1,025

Table 20. Bias and variability of recoveries for isotope-dilution standard compounds by water matrix type from ambient field samples collected in 2009–2010 and analyzed by using laboratory schedule 2434 or 4434.

[N, number of samples; NWIS, National Water Information System; NA, specific NWIS medium code not available for the water matrix type shown; RF_σ, relative F-pseudosigma; RSD, relative standard deviation; --, not applicable; NWIS medium codes are defined under the water matrix column; WWTP, wastewater treatment plant]

Isotope-dilution standard	NWIS medium code	Water matrix	Laboratory schedule	N	Mean (percent)	RSD (percent)	Median (percent)	RF_σ (percent)	Minimum (percent)	Maximum (percent)
16-Epiestriol-d_2	WB	Blended, untreated water supply	2434	11	77.2	7.3	78.4	6.1	69.3	89.4
	WE	Effluent, not landfill	2434	11	79.9	15.7	81.8	15.9	60.0	100
	WG	Groundwater	2434	47	54.2	48.1	69.1	49.5	13.7	89.2
	WS	Surface water	2434	6	78.1	13.4	78.1	15.1	65.7	90.2
	WT	Treated water supply	2434	11	73.9	6.9	73.7	5.9	65.5	81.9
	NA	WWTP influent[a]	2434	6	94.8	9.8	93.9	11.1	83.7	106
	WB	Blended, untreated water supply	4434	40	74.3	18.8	78.5	12.7	34.8	94.5
	WE	Effluent, not landfill	4434	48	76.7	27.1	80.7	13.4	23.8	117
	WG	Groundwater	4434	8	71.9	11.8	70.1	7.7	63.9	90.5
	NA	Hog manure slurry	4434	1	12.6	--	--	--	--	--
	WS	Surface water	4434	199	74.1	17.7	75.9	16.1	37.2	104
	WT	Treated water supply	4434	7	77.7	14.0	77.3	2.1	61.4	98.5
	NA	WWTP influent[a]	4434	13	79.1	47.5	69.1	53.7	24.3	154
17-*alpha*-Ethynylestradiol-d_4	WB	Blended, untreated water supply	2434	18	81.9	6.6	81.9	7.4	74.6	91.7
	WE	Effluent, not landfill	2434	14	85.5	14.7	85.2	17.2	65.8	104
	WG	Groundwater	2434	51	82.9	11.3	82.5	11.9	64.5	102
	WS	Surface water	2434	136	80.6	16.3	79.5	15.7	56.5	115
	WT	Treated water supply	2434	18	87.1	12.3	82.8	5.2	76.5	111
	NA	WWTP influent	2434	8	88.4	13.6	90.9	7.6	66.8	103
	WB	Blended, untreated water supply	4434	73	83.1	16.6	80.1	10.0	51.8	128
	WE	Effluent, not landfill	4434	151	81.6	20.1	79.8	15.9	41.4	135
	WG	Groundwater	4434	27	70.5	14.4	72.5	11.0	48.3	90.3
	NA	Hog manure slurry	4434	3	65.6	13.9	66.9	10.0	55.9	74.1
	NA	Hog manure spill impacted stream	4434	16	77.9	14.1	79.5	11.9	58.7	96.6
	SL	Sludge	4434	3	48.1	21.8	45.7	16.6	39.0	59.5
	WS	Surface water	4434	236	77.3	15.8	79.1	13.6	48.2	119
	WT	Treated water supply	4434	14	86.1	16.8	89.7	18.0	51.3	103
	NA	WWTP influent	4434	52	85.6	27.1	79.7	35.6	40.4	148
17-*beta*-Estradiol-$^{13}C_6$	WB	Blended, untreated water supply	2434	11	81.7	7.0	82.0	4.9	73.4	94.2
	WE	Effluent, not landfill	2434	11	82.0	9.1	83.2	5.7	69.9	94.0
	WG	Groundwater	2434	47	86.3	12.7	85.3	17.2	68.6	105
	WS	Surface water	2434	6	77.9	3.0	78.3	2.7	74.4	80.5
	WT	Treated water supply	2434	11	78.3	3.4	78.0	3.2	73.8	83.0
	NA	WWTP influent	2434	6	86.1	5.0	85.9	4.1	79.0	90.7
	WB	Blended, untreated water supply	4434	40	77.4	6.2	77.8	3.5	62.6	88.3
	WE	Effluent, not landfill	4434	48	83.5	14.0	82.2	11.5	61.5	111
	WG	Groundwater	4434	8	76.0	17.0	75.0	16.6	57.7	92.0
	NA	Hog manure slurry	4434	1	2.4	--	--	--	--	--
	WS	Surface water	4434	199	71.1	18.9	73.8	21.1	38.2	98.4
	WT	Treated water supply	4434	7	79.4	13.7	77.7	10.5	60.6	93.7
	NA	WWTP influent	4434	13	70.1	29.5	64.2	27.3	45.8	118

Table 20. Bias and variability of recoveries for isotope-dilution standard compounds by water matrix type from ambient field samples collected in 2009–2010 and analyzed by using laboratory schedule 2434 or 4434—Continued

[N, number of samples; NWIS, National Water Information System; NA, specific NWIS medium code not available for the water matrix type shown; RF_σ, relative F-pseudosigma; RSD, relative standard deviation; --, not applicable; NWIS medium codes are defined under the water matrix column; WWTP, wastewater treatment plant]

Isotope-dilution standard	NWIS medium code	Water matrix	Laboratory schedule	N	Mean (percent)	RSD (percent)	Median (percent)	RF_σ (percent)	Minimum (percent)	Maximum (percent)
Bisphenol A-d_{16}	WB	Blended, untreated water supply	2434	18	88.3	8.0	89.5	10.1	77.2	99.9
	WE	Effluent, not landfill	2434	14	93.0	15.0	95.9	13.9	65.1	111
	WG	Groundwater	2434	51	78.7	25.3	84.2	27.7	32.6	108
	WS	Surface water	2434	136	83.8	17.5	82.7	17.7	32.3	123
	WT	Treated water supply	2434	18	95.3	7.2	94.0	7.5	84.4	109
	NA	WWTP influent	2434	8	95.7	23.8	104	14.8	59.6	116
	WB	Blended, untreated water supply	4434	73	91.9	17.7	91.5	12.6	48.0	132
	WE	Effluent, not landfill	4434	151	86.4	22.6	85.6	24.7	43.3	151
	WG	Groundwater	4434	27	78.0	18.4	79.2	10.3	52.3	125
	NA	Hog manure slurry	4434	3	81.3	61.2	105	22.6	24.1	114
	NA	Hog manure spill impacted stream	4434	16	102	14.9	97.1	19.5	81.3	127
	SL	Sludge	4434	3	23.1	124	8.6	222	4.5	56.1
	WS	Surface water	4434	236	86.2	19.0	88.8	22.6	53.8	145
	WT	Treated water supply	4434	14	94.3	14.3	95.4	7.5	61.6	123
	NA	WWTP influent	4434	52	53.0	72.4	55.2	79.9	0.2	156
Cholesterol-d_7	WB	Blended, untreated water supply	2434	18	73.5	8.1	74.6	8.0	62.4	84.1
	WE	Effluent, not landfill	2434	14	75.4	11.6	75.5	12.5	60.2	88.4
	WG	Groundwater	2434	51	72.3	17.4	71.6	19.9	51.4	94.7
	WS	Surface water	2434	136	73.4	10.3	73.0	10.3	59.4	93.5
	WT	Treated water supply	2434	18	72.7	7.8	73.3	8.6	62.3	82.6
	NA	WWTP influent	2434	8	67.5	18.6	66.7	17.2	49.0	83.6
	WB	Blended, untreated water supply	4434	73	73.7	10.1	74.1	7.9	57.5	88.7
	WE	Effluent, not landfill	4434	151	69.1	17.7	69.5	17.5	41.7	100
	WG	Groundwater	4434	27	68.6	11.2	68.9	8.0	47.2	81.9
	NA	Hog manure slurry	4434	3	31.3	43.2	37.8	24.2	15.8	40.5
	NA	Hog manure spill impacted stream	4434	16	67.1	16.1	66.7	16.8	48.4	84.3
	SL	Sludge	4434	3	28.8	51.8	25.7	42.3	15.7	45.0
	WS	Surface water	4434	236	73.9	12.9	75.6	12.3	48.2	108
	WT	Treated water supply	4434	14	71.2	10.7	73.3	9.1	57.2	85.6
	NA	WWTP influent	4434	52	51.1	39.6	51.6	51.4	17.7	103
Diethylstilbesterol-d_8	WB	Blended, untreated water supply	2434	18	52.9	23.2	51.6	20.9	30.0	75.5
	WE	Effluent, not landfill	2434	14	77.0	10.9	75.2	11.1	64.4	94.6
	WG	Groundwater	2434	51	72.5	15.8	71.3	14.5	46.6	101
	WS	Surface water	2434	136	53.6	25.2	54.0	24.8	20.3	89.7
	WT	Treated water supply	2434	18	71.4	14.6	69.4	18.8	56.1	90.3
	NA	WWTP influent	2434	8	82.6	16.8	88.9	8.8	56.6	95.1
	WB	Blended, untreated water supply	4434	73	41.5	29.8	40.4	32.5	19.0	80.6
	WE	Effluent, not landfill	4434	151	70.8	24.6	68.9	23.3	13.5	109
	WG	Groundwater	4434	27	58.5	21.6	58.1	15.4	33.7	99.2
	NA	Hog manure slurry	4434	3	63.8	50.1	53.8	42.4	38.1	99.6
	NA	Hog manure spill impacted stream	4434	16	44.2	58.2	37.0	66.2	16.0	92.4
	SL	Sludge	4434	3	39.1	98.2	31.7	88.6	4.9	80.6
	WS	Surface water	4434	236	49.3	34.9	47.8	40.6	16.1	85.2
	WT	Treated water supply	4434	14	65.9	19.8	63.6	25.5	47.4	85.5
	NA	WWTP influent	4434	52	74.5	30.8	71.6	23.7	25.5	134

Table 20. Bias and variability of recoveries for isotope-dilution standard compounds by water matrix type from ambient field samples collected in 2009–2010 and analyzed by using laboratory schedule 2434 or 4434.—Continued

[N, number of samples; NWIS, National Water Information System; NA, specific NWIS medium code not available for the water matrix type shown; RF$_\sigma$, relative F-pseudosigma; RSD, relative standard deviation; --, not applicable; NWIS medium codes are defined under the water matrix column; WWTP, wastewater treatment plant]

Isotope-dilution standard	NWIS medium code	Water matrix	Laboratory schedule	N	Mean (percent)	RSD (percent)	Median (percent)	RF$_\sigma$ (percent)	Minimum (percent)	Maximum (percent)
Estrone-$^{13}C_6$	WB	Blended, untreated water supply	2434	11	85.8	10.3	86.0	8.6	71.8	103
	WE	Effluent, not landfill	2434	11	87.6	11.3	90.6	10.3	70.6	101
	WG	Groundwater	2434	47	84.0	21.0	80.5	17.8	0.0	109
	WS	Surface water	2434	6	82.9	2.7	82.9	1.9	79.8	86.4
	WT	Treated water supply	2434	11	84.5	6.2	84.4	2.8	75.2	94.2
	NA	WWTP influent	2434	6	95.4	11.1	93.2	12.6	85.0	111
	WB	Blended, untreated water supply	4434	40	79.3	6.5	79.4	5.3	61.3	93.9
	WE	Effluent, not landfill	4434	48	89.0	27.0	89.7	17.9	0.0	127
	WG	Groundwater	4434	8	74.5	27.5	78.2	14.7	34.1	97.5
	NA	Hog manure slurry	4434	1	87.3	--	--	--	--	--
	WS	Surface water	4434	199	77.3	16.0	78.7	13.0	49.6	104
	WT	Treated water supply	4434	7	80.2	19.0	78.5	11.8	53.1	100
	NA	WWTP influent	4434	13	51.7	48.7	58.6	30.9	4.0	93.4
Medroxyprogesterone-d_3	WB	Blended, untreated water supply	2434	11	84.3	7.9	83.6	9.7	75.6	92.8
	WE	Effluent, not landfill	2434	11	88.6	19.3	92.1	14.5	63.9	121
	WG	Groundwater	2434	47	67.0	33.4	70.8	27.9	7.2	114
	WS	Surface water	2434	6	86.6	14.1	86.0	16.9	74.1	102
	WT	Treated water supply	2434	11	84.6	9.9	81.8	8.6	76.4	102
	NA	WWTP influent	2434	6	104	21.7	104	23.5	77.4	134
	WB	Blended, untreated water supply	4434	40	47.6	50.2	45.9	56.7	0.7	94.8
	WE	Effluent, not landfill	4434	48	88.1	26.1	89.7	20.0	18.8	128
	WG	Groundwater	4434	8	75.9	21.3	78.7	17.1	48.5	95.1
	NA	Hog manure slurry	4434	1	12.9	--	--	--	--	--
	WS	Surface water	4434	199	37.9	65.7	36.0	86.3	0.0	114
	WT	Treated water supply	4434	7	77.9	19.3	76.4	10.4	50.2	98.5
	NA	WWTP influent	4434	13	30.2	95.0	19.0	172	0.5	83.0
Mestranol-d_4	WB	Blended, untreated water supply	2434	18	82.2	7.0	80.2	5.7	76.1	95.2
	WE	Effluent, not landfill	2434	14	80.6	10.0	83.3	10.4	64.4	91.2
	WG	Groundwater	2434	51	81.0	9.7	80.8	12.3	65.0	95.0
	WS	Surface water	2434	136	78.6	13.1	80.8	13.7	56.1	104
	WT	Treated water supply	2434	18	80.9	6.0	79.6	5.7	75.0	91.8
	NA	WWTP influent	2434	8	85.8	11.5	86.8	8.5	68.4	97.5
	WB	Blended, untreated water supply	4434	73	80.3	10.6	78.9	9.1	56.1	105
	WE	Effluent, not landfill	4434	151	78.1	17.7	77.5	13.1	45.5	125
	WG	Groundwater	4434	27	72.2	12.7	72.0	11.7	51.6	85.6
	NA	Hog manure slurry	4434	3	61.1	34.4	51.2	27.7	46.9	85.2
	NA	Hog manure spill impacted stream	4434	16	77.2	9.0	79.0	9.0	64.3	87.7
	SL	Sludge	4434	3	49.9	34.1	56.6	21.0	30.6	62.6
	WS	Surface water	4434	236	76.3	14.9	77.8	12.4	50.5	125
	WT	Treated water supply	4434	14	79.2	13.7	84.0	10.9	49.5	91.7
	NA	WWTP influent	4434	52	85.2	26.5	82.4	24.1	50.0	166

Table 20. Bias and variability of recoveries for isotope-dilution standard compounds by water matrix type from ambient field samples collected in 2009–2010 and analyzed by using laboratory schedule 2434 or 4434.—Continued

[N, number of samples; NWIS, National Water Information System; NA, specific NWIS medium code not available for the water matrix type shown; RF$_σ$, relative F-pseudosigma; RSD, relative standard deviation; --, not applicable. NWIS medium codes are defined under the water matrix column; WWTP, wastewater treatment plant]

Isotope-dilution standard	NWIS medium code	Water matrix	Laboratory schedule	N	Mean (percent)	RSD (percent)	Median (percent)	RF$_σ$ (percent)	Minimum (percent)	Maximum (percent)
Nandrolone-d_3	WB	Blended, untreated water supply	2434	11	85.8	11.2	83.1	12.3	70.9	102
	WE	Effluent, not landfill	2434	11	93.8	11.3	92.5	12.0	77.1	110
	WC	Groundwater	2434	47	74.4	27.5	78.7	27.7	1.6	107
	WS	Surface water	2434	6	81.6	6.1	82.1	3.4	73.0	87.6
	WT	Treated water supply	2434	11	86.1	5.8	84.9	3.9	78.6	94.8
	NA	WWTP influent	2434	6	98.4	17.1	92.0	13.5	79.1	125
	WE	Blended, untreated water supply	4434	40	66.3	19.3	68.3	14.0	21.7	85.6
	WE	Effluent, not landfill	4434	48	91.4	31.2	89.3	25.0	0.6	147
	WC	Groundwater	4434	8	77.1	30.4	84.5	10.4	30.3	104
	NA	Hog manure slurry	4434	1	0.7	--	--	--	--	--
	WS	Surface water	4434	199	61.4	26.0	62.6	31.9	7.1	92.5
	WT	Treated water supply	4434	7	83.0	19.6	84.9	8.4	49.9	101
	NA	WWTP influent	4434	13	672	66.3	64.4	49.4	5.1	170
					(milliliters)	(percent)	(milliliters)	(percent)	(milliliters)	(milliliters)
Sample volume	WE	Blended, untreated water supply	2434	18	441	4.4	442	5.3	413	471
	WE	Effluent, not landfill	2434	14	443	3.7	444	5.2	414	464
	WC	Groundwater	2434	51	453	5.9	449	2.7	416	589
	WS	Surface water	2434	136	432	8.1	436	7.7	283	499
	WT	Treated water supply	2434	18	437	4.3	430	5.7	410	468
	NA	WWTP influent	2434	8	426	4.3	430	3.3	396	450
	WE	Blended, untreated water supply	4434	73	702	8.0	699	7.2	423	815
	WE	Effluent, not landfill	4434	151	601	27.2	666	36.0	315	1,025
	WC	Groundwater	4434	27	476	5.9	474	3.7	453	604
	NA	Hog manure slurry	4434	3	172	127	47.9	293	44.4	424
	NA	Hog manure spill impacted stream	4434	16	458	3.7	460	3.6	432	486
	SL	Sludge	4434	3	180	103	106	121	43.1	391
	WS	Surface water	4434	236	594	36.1	476	56.8	242	973
	WT	Treated water supply	4434	14	471	2.3	474	1.6	447	485
	NA	WWTP influent	4434	52	563	26.5	627	32.9	314	817

[a]Includes sample matrices identified as WWTP influent or primary WWTP effluent.

of the procedure. Mean component recoveries relative to the total determined in each component were calculated using compound concentrations determined by quantitation relative to the injection internal standard compounds instead of by using the IDQ procedure and are not directly comparable to recoveries in other tables in this report. Mean component recoveries for cholesterol-d_7, cholesterol, and 3β-coprostanol only are shown in table 21. Tested components included:

(1) *Standard disk elution*: this was the standard 40-mL methanol elution of the GFF/C_{18} disk following sample extraction as described in section 9.8. Mean component recoveries for cholesterol-d_7, cholesterol, and 3β-coprostanol ranged from 34 to 45 percent, whereas recoveries for the other 17 analytes and 11 IDS tested were 97 percent or greater in this standard disk elution volume.

(2) *Sample filtrate*: this was the extracted water filtrate that passed through the GFF/C_{18} disk. The filtrate was extracted with dichloromethane using liquid-liquid extraction and processed with the other components through analysis. Mean recoveries for cholesterol-d_7, cholesterol, and 3β-coprostanol ranged from 50 to 63 percent, revealing that these three compounds were poorly retained by the GFF/C_{18} disk when extracting method compounds from reagent water. No more than 2 percent of any other analyte or IDS was observed in the filtrate, demonstrating excellent extraction efficiency from reagent water by the GFF/C_{18} disk.

(3) *GFF/C_{18} disk wash*: this was the 10 mL of 25-percent methanol in water mixture that is passed through the GFF/C_{18} disk as a disk wash step following completion of sample extraction as described in section 9.7.9. The GFF/C_{18} disk wash was evaporated to dryness and processed with the other test components through analysis. It contained no analytes or IDS compounds at recoveries greater than 1 percent (cholesterol the most), and demonstrated that this step, which removes some unwanted coextracted matrix before the final disk elution step, does not reduce recoveries.

(4) *Dichloromethane (DCM) disk elution*: this was an additional elution of the GFF/C_{18} disk following the standard disk elution step, and consisted of 20 mL of DCM. It was used to determine if a portion of the analytes were not completely eluted by the standard 40 mL of methanol or if an additional or alternative strong solvent was required to provide complete compound elution. Recoveries in the DCM disk elution component replicates did not exceed 3.4 percent (cholesterol-d_7; mean 2.2) for any analyte or IDS compound, indicating that the standard disk elution step was sufficient to achieve high recovery of the method compounds from the disk.

(5) *Bottle rinse*: after decanting the sample water into the extractor, the glass sample bottle was rinsed with 20 mL of DCM to check for compounds that adhere to the glass surface. This might be expected for lower solubility compounds, such as cholesterol and 3β-coprostanol, and has been observed for pesticides with low water solubility (such as dichlorodiphenyltrichloroethane [DDT] and permethrin) (Foreman and Foster, 1991). This bottle rinse component contained no more than 3 percent (again cholesterol-d_7; mean 1.3) of any analyte or IDS compound.

One possible explanation for the substantial sorbent breakthrough by the sterols might be related in part to their water solubility. Based on partition theory, compounds like cholesterol and 3β-coprostanol that have lower water solubility compared to the other method analytes are predicted to partition strongly from water to solid sorbents such as C_{18}, and, likewise, to suspended particles, colloids, and dissolved organic matter. Thus, good isolation (high sorbent retention and little sorbent breakthrough) on C_{18} of these sterols during extraction of spiked reagent water might be expected. However, partitioning from water to any sorbent requires that the compound be truly "dissolved" in the water. The substantial amount of sorbent breakthrough by the sterols during SPE of non-salted reagent water indicates that the cholesterol and 3β-coprostanol might not be completely dissolved. If a substantial portion of these

Table 21.   Mean and standard deviation of relative absolute recovery from selected extraction-related components examined to evaluate losses of cholesterol-d_7, cholesterol, and 3-*beta*-coprostanol during solid-phase extraction of unsalted reagent water fortified with method analytes.[a]

[DCM, dichloromethane; MeOH, methanol; disk, includes the combined glass-fiber filter (GFF) and octadecylsilyl (C_{18}) silica solid-phase extraction disk; mL, milliliters; ±, plus or minus]

Analyte	Standard disk elution[b] (40-mL MeOH elution of the disk)	Sample filtrate (sample water that passed through disk)	Disk wash (10-mL wash of disk with 25-percent MeOH in water solution]	DCM disk elution (extra 20-mL DCM elution of the disk)	Bottle rinse (DCM rinse of glass sample bottle)
			(percent)		
Cholesterol-d_7	45.4 ± 22.8	50.5 ± 25.4	0.6 ± 0.2	2.2 ± 1.6	1.3 ± 1.6
Cholesterol	34.2 ± 18.0	62.6 ± 20.1	0.6 ± 0.4	1.7 ± 1.2	0.9 ± 1.0
3-*beta*-Coprostanol	34.4 ± 16.8	62.9 ± 18.0	0.6 ± 0.3	1.6 ± 1.1	0.5 ± 0.4

[a]Triplicate 500-mL reagent-water samples contained in amber glass bottles were fortified with the method analytes and extracted using standard conditions. No sodium chloride (salt) was added to the water prior to extraction. Absolute recoveries are relative to the total mass recovered for all five measured components. See "Cholesterol-d_7 Recoveries in Non-salted Regaent Water" section for details.

[b]Relative recoveries for the other method analytes and isotope dilution standard compounds were 97 percent or greater in this standard disk elution volume.

analytes partitions to the small amount of DOC (typical DOC concentration is less than 0.016 mg/L in the Solution 2000 reagent water) or to any fine particulate matter in the reagent water, or both, then this non-dissolved portion will not be available to partition to the C_{18} sorbent and also might not be physically "captured" by either the C_{18} disk matrix or the GFF that overlays the disk.

Indeed, Landrum and Giesy (1981) observed substantial breakthrough on an Amberlite™ XAD-4 sorbent column of benzo(a)pyrene, a compound with very low water solubility, when spiked into reagent water, and breakthrough was worse when DOC was added to the water because this compound partitions to the DOC that is not well retained by the sorbent. De Llasera and others (2007) also noted increasing breakthrough of moderately polar pesticides on C_{18} sorbent with increasing sample DOC concentration. Foreman and Foster (1991) observed 14-percent breakthrough using stacked 10-g C_{18} SPE sorbent columns for DDT in reagent water, whereas pesticides with somewhat higher water solubilities (for example, dichlorodiphenyldichloroethene (DDE) and atrazine) exhibited no breakthrough. Based on these findings, samples with high concentrations of DOC or fine colloids might be expected to lead to incomplete isolation (breakthrough) on C_{18} of the sterols (and possibly other method analytes) during extraction. Indeed, somewhat lower and more variable cholesterol-d_7 recoveries are observed in matrices with higher suspended solid and DOC concentrations (WWTP influent and primary effluent, sludge, hog manure slurry samples) compared with other matrices (tables 12–14, and 20).

Nevertheless, cholesterol-d_7 recoveries even in these complex matrices (salted or not) usually were substantially greater than the recoveries obtained from non-salted reagent water. The addition of NaCl to reagent water appears inconsequential to recoveries of the other method analytes based on IDS recoveries before and after salt use was implemented (January, 2009). The actual mechanism underlying the improved sorbent retention of the sterols from reagent water containing salt has not been elucidated. Our evaluation of NaCl addition to reagent-water to improve recoveries of the sterols was prompted by the improvement in recoveries obtained for selected analytes (for example, prometon in LS 2001 and 2033) in salted versus unsalted laboratory reagent-water spikes; although in those lab schedules, salt is added only to the LRS and LRB samples and not the field samples (NWQL SOP ORGP0053.x, "Sample preparation of 1-liter filtered water samples by C-18 solid-phase extraction (Method 2001)" [Stewart and others, National Water Quality Laboratory, written commun., 2009]).

Assessment of Blank Contamination and Determination of Detection and Reporting Levels

As noted in section 4 ("Interferences and Sample Contamination"), inadvertent contamination of samples with method analytes might occur because (1) many of them are common biogenic compounds, (2) some are used pharmaceutically, and, (3) in the case of bisphenol A, it is a

chemical used to make polycarbonate plastic and epoxy resins that are used in a diverse array of products or applications. The scope of this contamination potential is monitored by using laboratory- and field-blank samples as described in section 13 ("Quality Assurance/Quality Control") and summarized in this section of the report. In addition, considerations of blanks are an integral component in the determination of detection levels, especially for blank-limited analytes, and detection-level determinations also are addressed in this section. Blank data also are evaluated with respect to the detection and reporting levels applied to laboratory and field data presented in this report.

Blank-Limited Analytes

Bisphenol A, cholesterol, and 3β-coprostanol were detected in all laboratory reagent-water blank and field-blank samples, and data for these three blank-limited analytes are reported to NWIS using the minimum reporting level (MRL) convention (Childress and others, 1999). Table 22 summarizes concentration data for these three analytes for laboratory reagent-water blank samples and HDPE bottle lot-check samples (prepared identical to LRBs) from 2009, LRB samples from 2010, and the combined data for 2009–2010.

These blank data were used to estimate detection levels for the blank-limited analytes by using the standard deviation of the blank sample data in the simple parametric-based detection estimate equation described in the USEPA's method detection limit (MDL) procedure (U.S. Environmental Protection Agency, 1986), where MDL is calculated as:

$$MDL = s \times t \tag{11}$$

where

s	=	standard deviation of the determined concentrations, in nanograms per liter
t	=	Student's t-value at the 99-percent confidence level (alpha = 0.01) and N minus one degrees of freedom, where N is the number of samples.

The USEPA's MDL is defined (U.S. Environmental Protection Agency, 1986, page 1 of Appendix B to Part 136) as "…the minimum concentration of a substance that can be measured and reported with 99% (percent) confidence that the analyte concentration is greater than zero and is determined from analysis of a sample in a given matrix containing the analyte." The USEPA's MDL procedure estimates the analyte's MDL concentration (1) using the standard deviation of concentration data obtained from replicate samples ($N \geq 7$) that are fortified with the analyte at a concentration that is within five times the determined MDL and (2) by assuming that the resultant variation from these (typically few) replicates is adequately represented by a parametric (Student's t) distribution that is centered at zero concentration. For analytes with frequent detections in blanks, the use of variation data obtained from blank replicates provides a more direct (no spiking required) and accurate determination of the detection level because it better represents the "true" variation in the blank distribution in comparison with an assumed variation derived from low-concentration spike samples that are used to calculated the MDL.

More importantly, analytes that are blank limited commonly have blank-concentration distributions that are not centered on zero, but are offset to a higher concentration. Also shown in table 22 are "mean-offset MDLs" that are calculated by adding the mean laboratory blank concentration to the MDL calculated using equation 11. The mean-offset MDL accounts for the fact that the blank distribution is not centered on zero concentration and, thus, represents the estimated minimum concentration of a substance that can be measured and reported with 99-percent confidence that the analyte concentration in the sample is greater than that in the blanks.

Bisphenol A

Bisphenol A concentrations in 2009 blanks were greater, and considerably more variable, than those in 2010 blanks, indicating a possible reduction in BPA contamination sources beginning in 2010. This was possibly attributable to minor method changes including implementation of substitute IDS compounds; however, the source of bisphenol A contamination has not been fully elucidated. Estimated MDLs and mean-offset MDLs for BPA based on the standard deviation of 2009, 2010, and combined blank sample data are shown in table 22.

Based on sparse prior blank data, an MRL of 100 ng/L was applied to samples prepared and analyzed in 2009. The mean-offset MDL calculated from 2009 laboratory blank data was less than this 100 ng/L censoring limit. Nevertheless, BPA's MRL applied to samples prepared and analyzed in 2010 was raised to 200 ng/L as a precaution against false positives based on the magnitude of the 99th- percentile concentration (172 ng/L) from the 2009 blanks. Note: The 99th-percentile concentration (or the highest determined concentration of tested blank samples that is less than the calculated 99th-percentile concentration) of blind or method blank measurements sometimes is used as the long-term method detection level concentration in select NWQL methods (see Appendix C in U.S. Geological Survey, 2010). Based on 2010 laboratory blank data, the MRL for BPA was lowered to 100 ng/L on October 1, 2011. This MRL is 10 times greater than the 2010 mean-offset MDL of 10 ng/L, but is being applied as a conservative reporting level partly in consideration of field-blank data for BPA (as described in the "Laboratory Reagent-Water Blank (LRB) and Field Blank Sample Results" section that follows).

Cholesterol and 3-*beta*-Coprostanol

Mean (or median) concentrations for 3β-coprostanol were similar for 2009 and 2010; likewise was the case for cholesterol (table 22). Limited testing revealed that these two sterols are introduced to LRB sample extracts primarily during the post-extraction evaporation steps, likely from introduction of dust particles (Weschler and others, 2011) during the three solvent-evaporation steps. Based on prior blank data for cholesterol and 3β-coprostanol, a conservative MRL value of 2,000 ng/L for each sterol was applied to samples prepared and analyzed in 2009. Very low cholesterol-d_7 recoveries in non-salted LRBs analyzed before 2009 resulted in a high bias artifact in sterol concentrations in the LRBs because the IDS and sterol analytes did not emulate each other during sample preparation. The two sterols contaminate the extract after the extraction step, whereas the cholesterol-d_7 loss was during the extraction step. Data collected in 2009 showed dramatically lower cholesterol or 3β-coprostanol concentrations in the LRBs (table 22), which was a direct consequence of the much improved cholesterol-d_7 recovery achieved during the GFF/C$_{18}$ disk SPE isolation step because of NaCl addition to the LRB (and all) samples before extraction. Thus, the MRLs for these two sterols were lowered to 200 ng/L for samples analyzed starting in 2010.

11-Ketotestosterone

Tetradecamethylcycloheptasiloxane ($C_{14}H_{42}Si_7O_7$) with a mass-to-charge ratio (m/z) of 518.1315 is thought to be the compound that interferes with determination of the derivatized 11-ketotestosterone at near instrument detection level concentrations. This "siloxane" compound is believed to come from thermal decomposition of the polydimethylsiloxane stationary phase of the gas chromatographic capillary column or possibly from the silicone septum used for the GC or reaction vials. This compound also is present in various personal care, household, and other silicone containing products (see, for example, Horii and Kannan, 2008, and references therein). In the ion source, this siloxane compound can lose a methyl (CH_3) group to form an ion (m/z = 503.1081) that, if present, produces some positive signal bias in the precursor ion used for 11-ketotestosterone (m/z = 503.3) under the low-resolution tandem mass-spectrometry (MS/MS) conditions used in this method. The level of resulting interference in the monitored 11-ketotestosterone product quantitation ion (m/z = 323.2) averages about 0.3 ng/L, with a calculated mean-offset MDL of about 0.8 ng/L (maxiumum of about 1.2 ng/L) as established from analysis of 194 LRB and bottle lot-check blank samples (table 23). Therefore, a MRL of 2 ng/L is used for 11-ketotestosterone. Alternative precursor ions (table 6) for the determination of 11-ketotestosterone are at least 10 times less responsive compared to the m/z 503.3 ion.

Laboratory Reagent-Water Blank and Field-Blank Sample Results

The concentrations of analytes detected in LRB samples and in field-blank samples submitted for LS 2434 and 4434 are summarized in table 24, along with the numbers of detections relative to the detection and reporting levels applied to samples in 2009–2010 or 2010 only for those analytes affected by the IDS substitutions implemented in January 2010

Table 22. Statistical summary of concentrations and estimated method detection limits for bisphenol A, cholesterol, and 3-*beta*-coprostanol from laboratory reagent-water blank (LRB) samples and bottle lot-check samples from 2009 and LRB samples from 2010.

[MDL, method detection limit; MRL, minimum reporting level; N, total number of blank samples; NA, not applicable; n≥/L, nanograms per liter]

Analyte	Year	N	Mean (ng/L)	Standard deviation [s] (ng/L)	RSD (percent)	Concentration in LRB samples					MDL (ng/L)	Mean-offset MDL[a] (ng/L)	MRL applied (ng/L)
						Minimum (ng/L)	Median (ng/L)	95th percentile (ng/L)	99th percentile (ng/L)	Maximum (ng/L)			
Bisphenol A	2009	129	15.9	31.5	199	0.0	5.3	65.4	172	208	74	90	100
	2010	52	4.0	2.6	66	1.6	3.1	10.4	12.4	13.6	6.3	**10**	200
	2009–10	181	12.5	27.1	218	0.0	4.0	53.6	155	208	54	76	NA
Cholesterol	2009	124	30.5	14.6	48	0.0	28.3	54.1	74.1	90.9	34	65	2,000
	2010	52	36.0	27.6	77	1.3	30.3	61.1	147	198	56	**102**	200
	2009–10	176	32.1	19.5	61	0.0	28.9	60.8	92.9	198	46	78	NA
3-*beta*-Coprostanol	2009	124	14.8	10.8	73	0.0	19.8	26.5	28.7	28.9	26	40	2,000
	2010	52	19.1	21.8	114	2.7	14.4	34.4	117	135	52	**71**	200
	2009–10	176	16.1	15.0	93	0.0	18.5	27.7	55.7	135	35	51	NA

[a]Mean-offset MDL = Mean + MDL. The value in **bold** from 2010 was used as the applied detection level value in table 25.

Table 23.   Summary of estimated concentrations and calculated method detection limit (MDL) and minimum reporting level for 11-ketotestosterone based on response from interfering tetradecamethylcycloheptasiloxane peak in 194 laboratory reagent-water blank samples and bottle lot-check blank samples from 2009–2010.

[ng/L, nanograms per liter]

	Tetradecamethyl-cycloheptasiloxane[a]
Number of blank samples [N]	194
Estimated blank concentration	
Mean concentration in blanks, ng/L	0.29
Standard deviation [s], ng/L	0.20
Relative standard deviation, percent	69.6
Minimum, ng/L	0.004
Median, ng/L	0.31
95th percentile, ng/L	0.58
99th percentile, ng/L	0.91
Maximum, ng/L	1.15
Estimated detection and reporting levels	
Student's t-value[b]	2.35
Method detection limit (MDL = $s \times t$), ng/L	0.48
Mean-offset MDL (= Mean + MDL), ng/L	0.77
Minimum reporting level, ng/L	2

[a]Chemical Abstract Service Registry Number 107–50–6. Molecular weight: 518 1315 atomic mass units.

[b]Student's t-value for 99-percent probability ($\alpha = 0.01$) and N-1 degrees of freedom.

(see section 10.7). Summarized results for BPA, cholesterol, and 3β-coprostanol include non-reported (to NWIS) detections that were less than the MRL applied at the time, and are included to show concentrations relative to the MRL and the mean-offset MDL (for 2010) shown in table 22. The LRB concentrations for BPA, cholesterol, and 3β-coprostanol shown in table 24 do not match those presented in table 22. The data in table 24 are for the LRBs only and do not include the HDPE bottle lot-check blank samples used in the summary presented in table 22 for these three blank-limited compounds. The LRB results in table 24 also include several high (relative to most LRB) cholesterol and 3β-coprostanol concentrations that were attributed to injection carryover from wastewater-treatment plant (WWTP) influent or primary effluent samples that were positioned before the LRB in the gas chromatography/tandem mass spectrometry (GC/MS/MS) analysis sequence. These few carryover incidences preceded the GC/MS/MS sequencing change strategy implemented to minimize injection carryover risk from WWTP influent or other matrices containing very high sterol concentrations (see section 10.6), and these anomalous high concentrations were omitted from the statistical summaries shown in table 22.

Although BPA, cholesterol, and 3β-coprostanol were "detected" in all or nearly all LRB and field-blank samples, only one field-blank value (a LS 2434 sample) for BPA and one LRB value for cholesterol exceeded the applied MRL censoring concentrations. Several precensored values for each

blank type do exceed the mean-offset MDL (from 2010 data; table 22). Comparative high BPA concentrations (443 ng/L maximum) for the field blanks submitted for LS 2434 (note smaller N) indicate that the extra field-filtration steps might be an additional source of BPA contamination compared to the LRBs and to field blanks submitted for LS 4434.

For the remaining analytes, either they were not detected in the blanks or they were detected infrequently; all but three of these detections were less than the applied detection level (or less than the MRL for 11-ketotestosterone). One LRB value for ethynylestradiol and two field-blank values (LS 4434) for *cis*-androsterone (one of which was greater than the applied reporting level) exceeded the applied detection level.

Analyte Detection Levels

Estimated detection levels for the 16 analytes that are not blank-limited were determined by using the multi-concentration spiking procedure in ASTM International's Standard Practice D6091–07 for the determination of the interlaboratory detection estimate (IDE; ASTM International, 2007). The IDE procedure has several key advantages in comparison to the USEPA MDL procedure (U.S. Environmental Protection Agency, 1986). In particular, the IDE procedure simplifies detection-estimate determinations for multi-analyte methods with varying analyte responses typical for many organic methods because the IDE is designed as a multi-concentration procedure. Thus, it does not require the cumbersome, successive iterative determinations of the MDL if the original (and subsequent, typically lower) spiking level used in the MDL procedure is not between 1 and 5 times the determined MDL value. In addition, the IDE procedure considers (through three model scenarios) changes in the standard deviation with concentration to determine several detection-related parameters, whereas the MDL procedure assumes a constant standard deviation from the spiking level down to zero concentration (see ASTM International, 2007, and references therein).

The IDE procedure and associated DQCALC software (ASTM International, 2010) are used to calculate the USEPA MDL value for each spiking level. Furthermore, the IDE procedure is used to calculate Currie's critical level (Lc) that, like the MDL, is the estimated concentration where the risk of false positives is predicted to be no more than 1 percent in the tested matrix (spiked reagent water for this report). Theoretically, Lc and MDL values will be nearly identical; however, they might differ because the IDE procedure uses a spike concentration relative to determined concentration (calibration-like) model to estimate Lc (and IDE). Furthermore, the IDE procedure performs the calculations assuming no change in standard deviation ("constant" model) or changes in standard deviation with concentration based on three model options: "straightline," "exponential," and a "hybrid" model developed by Rocke and Lorenzato (1995) (ASTM International, 2007).

Table 24. Summary of method analyte detections in laboratory reagent-water blank (LRB) and field-blank samples submitted for laboratory schedules 2434 (2434FB) and 4434 (4434FB) in 2009–2010 or 2010 only.[a]

[IRL, interim reporting level; DL, detection level; MRL, minimum reporting level; N, total number of samples; ng/L, nanograms per liter; RL, reporting level; --, not applicable; ≥, greater than or equal to]

Method analyte	Applied DL (ng/L)	RL in 2009/2010[b] (ng/L)	RL type	Blank type	N	Number of detections			Minimum (ng/L)	Median (ng/L)	95th percentile (ng/L)	99th percentile (ng/L)	Maximum (ng/L)
						All	≥DL	≥RL					
11-Ketotestosterone	1	2	MRL	LRB	115	0	--	--	--	--	--	--	--
				2434FB	15	0	--	--	--	--	--	--	--
				4434FB	70	0	--	--	--	--	--	--	--
17-*alpha*-Estradiol	0.4	0.8	IRL	LRB	115	5	0	0	0	0	0	0.11	0.12
				2434FB	15	0	--	--	--	--	--	--	--
				4434FB	70	0	--	--	--	--	--	--	--
17-*alpha*-Ethynylestradiol	0.4	0.8	IRL	LRB	115	3	1	0	0	0	0	0.17	0.57
				2434FB	15	0	--	--	--	--	--	--	--
				4434FB	70	0	--	--	--	--	--	--	--
17-*beta*-Estradiol	0.4	0.8	IRL	LRB	115	4	0	0	0	0	0	0.15	0.15
				2434FB	15	0	--	--	--	--	--	--	--
				4434FB	70	1	0	0	0	0	0	0.04	0.14
3-*beta*-Coprostanol	71	2,000/200	MRL	LRB	115	114	5	0	0	19.7	39	231	1,430
				2434FB	15	15	2	0	7.4	20.4	--	--	758
				4434FB	70	65	4	0	0	13.5	175	837	938
Androstenedione	0.4	0.8	IRL	LRB	52	0	--	--	--	--	--	--	--
				2434FB	4	0	--	--	--	--	--	--	--
				4434FB	22	0	--	--	--	--	--	--	--
Bisphenol A	10	100/200	MRL	LRB	115	114	11	0	0	2.9	13.6	24.8	30.0
				2434FB	15	15	4	1	0.8	3.8	--	--	443
				4434FB	70	70	7	0	0.6	3.2	15.7	39.8	53.1
Cholesterol	102	2,000/200	MRL	LRB	115	110	6	1	0	30.7	115	526	2,440
				2434FB	15	15	3	0	12.7	33.6	--	--	830
				4434FB	70	70	4	0	2.3	23.1	253	1,220	1,770
cis-Androsterone	0.4	0.8	IRL	LRB	115	2	0	0	0	0	0	0.09	0.15
				2434FB	15	0	--	--	--	--	--	--	--
				4434FB	70	1	1	1	0	0	0	0.46	1.48
Dihydrotestosterone	2	4	IRL	LRB	52	0	--	--	--	--	--	--	--
				2434FB	4	0	--	--	--	--	--	--	--
				4434FB	22	0	--	--	--	--	--	--	--
Epitestosterone	2	4	IRL	LRB	115	2	0	0	0	0	0	0.10	0.16
				2434FB	15	0	--	--	--	--	--	--	--
				4434FB	70	0	--	--	--	--	--	--	--
Equilenin	1	2	IRL	LRB	115	2	0	0	0	0	0	0.09	0.11
				2434FB	15	0	--	--	--	--	--	--	--
				4434FB	70	0	--	--	--	--	--	--	--

Table 24. Summary of method analyte detections in laboratory reagent-water blank (LRB) and field-blank samples submitted for laboratory schedules 2434 (2434FB) and 4434 (4434FB) in 2009–2010 or 2010 only[a].—Continued

[IRL, interim reporting level; DL, detection level; MRL, minimum reporting level; N, total number of samples; ng/L, nanograms per liter; RL, reporting level; --, not applicable; ≥, greater than or equal to]

Method analyte	Applied DL (ng/L)	RL in 2009/2010[b] (ng/L)	RL type	Blank type	N	Number of detections			Minimum (ng/L)	Median (ng/L)	95th percentile (ng/L)	99th percentile (ng/L)	Maximum (ng/L)
						All	≥DL	≥RL					
Equilin	2	4	IRL	LRB	115	1	0	0	0	0	0	0	0.83
				2434FB	15	0	--	--	--	--	--	--	--
				4434FB	70	0	--	--	--	--	--	--	--
Estriol	1	2	IRL	LRB	115	1	0	0	0	0	0	0	0.04
				2434FB	15	0	--	--	--	--	--	--	--
				4434FB	70	0	--	--	--	--	--	--	--
Estrone	0.4	0.8	IRL	LRB	52	0	--	--	--	--	--	--	--
				2434FB	4	0	--	--	--	--	--	--	--
				4434FB	22	0	--	--	--	--	--	--	--
Mestranol	0.4	0.8	IRL	LRB	115	2	0	0	0	0	0	0.06	0.11
				2434FB	15	0	--	--	--	--	--	--	--
				4434FB	70	0	--	--	--	--	--	--	--
Norethindrone	0.4	0.8	IRL	LRB	52	0	--	--	--	--	--	--	--
				2434FB	4	0	--	--	--	--	--	--	--
				4434FB	22	0	--	--	--	--	--	--	--
Progesterone	4	8	IRL	LRB	52	0	--	--	--	--	--	--	--
				2434FB	4	0	--	--	--	--	--	--	--
				4434FB	22	0	--	--	--	--	--	--	--
Testosterone	0.4	0.8	IRL	LRB	52	0	--	--	--	--	--	--	--
				2434FB	4	0	--	--	--	--	--	--	--
				4434FB	22	0	--	--	--	--	--	--	--
trans-Diethylstilbestrol	0.4	0.8	IRL	LRB	115	4	0	0	0	0	0	0.28	0.36
				2434FB	15	0	--	--	--	--	--	--	--
				4434FB	70	0	--	--	--	--	--	--	--

[a]The nonparametric statistics were calculated using all samples, including non-detections that were treated as zero values.

[b]If only one value is listed, then the reporting level was unchanged between 2009 and 2010.

The IDE is defined in Standard Practice D6091-07 as "the lowest concentration at which there is 90 percent confidence that a single measurement from a laboratory selected from the population of qualified laboratories represented in an interlaboratory study will have a true detection probability of at least 95 percent (5 percent false negative risk) and a true nondetection probability of at least 99 percent (1 percent false positive risk when measuring a blank sample)" (ASTM International, 2007). For the analytical method presented in this report, the IDE procedure was applied as, and the determined IDE value used as, an *intra*laboratory detection estimate. As such, the IDE should be approximately equal to a laboratory reporting level (LRL) value, which Childress and others (1999) note is calculated by multiplying the determined MDL (or long-term method detection level value) by a minimal factor of 2 (see additional discussion in Appendix C of Office of Water Quality Technical Memorandum 2010.07 (U.S. Geological Survey, 2010)).

Eight replicate reagent-water samples (about 450 mL) were spiked at about 0.1, 0.4, 1, 2, 3, and 4 ng/L of each analyte (BPA was 10 times higher, and cholesterol and 3β-coprostanol were 100 times higher at each level) as shown in table 25, and were fortified with the normal amount of the IDSs. Unspiked (0 ng/L) replicates also were included. All replicates were prepared and analyzed by the method in a manner identical to field samples in three independent sample preparation and analysis sets. A summary of some detection- and quantitation-related parameters calculated using the DQCALC software is shown in table 25. Results for the four blank-limited compounds (BPA, cholesterol, 3β-coprostanol, and 11-ketotesterone) are included in this summary for comparison with the blank-based mean-offset MDL estimates shown in tables 22 and 23.

For most analytes, there were no detections in the 0-ng/L (unspiked) replicates and no or very few detections in the 0.1-ng/L replicates; these concentration levels were omitted from calculations. For several analytes, the 0.4-ng/L level also was omitted because of no or few detections. One unusually high value at the 0.4-ng/L level was omitted for several analytes. The number of determined values used in the calculation at each spiking level is provided in table 25, along with the number of these values where the analyte did not meet secondary ion qualifying criteria. These "non-qualified" values were included in the calculation to provide at least four spiking levels for inclusion in the models, but are an indication of the concentration level where reliable qualitative "detection" might not always be high. All calculations are based on the determined concentration from the quantitation ion response that might be substantially greater than the GC/MS/MS response for the two secondary ions that are used to ensure qualitative identification of the analyte in samples.

Estimated Lc and IDE concentrations calculated using each of the four standard deviation models noted previously are shown in table 25, along with a notation of the model having the best "fit" parameters under the "Standard deviation model" column. Values of Lc shown in bold are similar from

two or more models and are considered reasonable estimates of Lc. Also shown is the calculated USEPA MDL at each spike level, along with a notation of whether the MDL is considered "valid" by the DQCALC software. The criteria for "validity" are a MDL concentration that is no greater than the corresponding spiking level but not less than 20 percent of the spike level, and a minimum of seven values at the spike level. (Note: the DQCALC software simply considers the number of values [including zero] in determining this "minimum," not the number of actual analyte "detections" that meet identification criteria).

The "valid" MDL value shown in bold in table 25 was similar to the bolded Lc values. Note that the determined MDL can vary substantially based on the spiking level, with generally increasing MDL as the spike concentration increases. This change in MDL value based on spike level highlights the requirement within the USEPA MDL procedure to perform iterative determinations of the MDL at successively decreasing spiking level to ensure that the determined MDL value is within 1 to 5 times the spike level as mentioned previously. This iterative process can result in a nearly equivalent number of total measurements to estimate the MDL value as is required by the IDE procedure; yet, the IDE procedure provides a more practical approach for determining detection levels, especially for multi-analyte methods that often have subtantially different instrument detector response characteristics.

Also shown in table 25 are the detection and reporting levels applied to the validation data in this report and to data for field samples prepared and analyzed beginning in 2010 (or earlier for select analytes). These interim detection and reporting levels initially were estimated using calibration and earlier performance data. Data for 16 analytes were reported using the LRL convention, with concentration data less than the detection level provided for this "information rich" mass spectrometry method (Childress and others, 1999). An interim reporting level (IRL) "type" code was used for all analytes except 11-ketotestosterone, bisphenol A, cholesterol, and 3β-coprostanol that are reported to NWIS using the MRL convention (Childress and others, 1999). The USEPA minimum level value, defined as 3.18 times the MDL, is only calculated by DQCALC when the MDL is deemed "valid," and is shown in table 25 for comparison with the IDE and the applied interim reporting level.

For many analytes, the Lc concentrations estimated using two or more of the standard deviation models were similar and also compared well with the lowest "valid" MDL concentration (values in bold in table 25). For 8 of the 16 analytes reported using the LRL convention, these Lc and corresponding valid MDL values agreed well with (that is, were within 1.5 times) the interim detection level concentrations applied to the performance data in this report. For the other 8 analytes, the Lc and MDL value were somewhat (more than 1.5 times) greater or less than the interim applied detection levels. An exception was progesterone, with Lc and MDL values estimated as 1 ng/L

Table 25.   Estimates of Currie's critical level (Lc), the *intra*laboratory detection estimate (IDE), and the U.S. Environmental Protection Agency (USEPA) method detection limit (MDL) and minimum level (ML) concentrations for method analytes using ASTM standards D6091–07 (ASTM International, 2007) and D7510–10 (ASTM International, 2010).

[N, no; ng/L, nanograms per liter; Y, yes; --, not applicable; values in **bold** considered to exhibit good agreement for Lc by different models and with the estimated MDL]

Method analyte	Standard deviation model[c]	Lc (ng/L)	IDE[d] (ng/L)	Spike level[e] (ng/L)	Standard deviation at spike level (ng/L)	Number of values[f]	Number of non-qualified values[g]	MDL (ng/L)	"Valid" MDL?[h]	ML[i] (ng/L)	Detection level (ng/L)	Reporting level (ng/L)
11-Ketotestosterone	Constant	**1.4**	2.5	0.4	0.32	7	4	1.0	N	--	0.8[i]	2
	Straight line	1.4	2.4	1.1	0.65	8	3	1.9	N	--		
	Exponential	1.2	2.2	2.2	0.70	8	0	**2.1**	Y	6.7		
	Hybrid	1.4	2.4	3.2	0.51	8	0	1.5	Y	4.9		
				4.3	0.45	8	0	1.4	Y	4.3		
17-*alpha*-Estradiol	Constant	1.1	1.8	0.4	0.36	8	1	1.1	N	--	0.4	0.8
	Straight line	**0.66**	1.3	1.1	0.25	8	0	**0.74**	Y	2.4		
	Exponential	**0.71**	1.3	2.2	0.40	8	0	1.2	Y	3.8		
	Hybrid	**0.76**	1.4	3.2	0.32	8	0	0.97	Y	3.1		
				4.3	0.61	8	0	1.8	Y	5.8		
17-*beta*-Estradiol	Constant	0.76	1.3	0.4	0.30	8	1	0.90	N	--	0.4	0.8
	Straight line	**0.37**	0.70	1.1	0.13	8	0	**0.38**	Y	1.2		
	Exponential	**0.38**	0.70	2.2	0.11	8	0	0.33	N	--		
	Hybrid	**0.45**	0.79	3.2	0.35	8	0	1.0	Y	3.3		
				4.3	0.43	8	0	1.3	Y	4.1		
17-*alpha*-Ethynylestradiol	Constant	0.94	1.6	0.4	0.17	7	1	0.54	N	--	0.4	0.8
	Straight line	**0.33**	0.68	1.1	0.18	8	0	**0.53**	Y	1.7		
	Exponential	**0.42**	0.80	2.2	0.32	8	0	0.96	Y	3.0		
	Hybrid	**0.45**	0.83	3.2	0.34	8	0	1.0	Y	3.2		
				4.3	0.50	8	0	1.5	Y	4.7		
3-*beta*-Coprostanol	Constant	48	82	0	9.2	8	0	**28**	N	--	71[j]	200
	Straight line	**21**	41	11	7.5	8	0	22	N	--		
	Exponential	**21**	38	43	2.6	8	0	7.9	N	--		
	Hybrid	**24**	42	108	6.2	8	0	19	N	--		
				217	14	8	0	42	N	--		
				325	23	8	0	69	Y	218		
				434	17	8	0	52	N	--		

Detection and reporting levels applied in this report[b]

Table 25. Estimates of Currie's critical level (Lc), the *intralaboratory* detection estimate (IDE), and the U.S. Environmental Protection Agency (USEPA) method detection limit (MDL) and minimum level (ML) concentrations for method analytes using ASTM standards D6091–07 (ASTM International, 2007) and D7510–10 (ASTM International, 2010).—Continued

[N, no; ng/L, nanograms per liter; Y, yes; –, not applicable; values in **bold** considered to exhibit good agreement for Lc by different models and with the estimated MDL]

Method analyte	Standard deviation model[c]	Lc (ng/L)	IDE[d] (ng/L)	Spike level[e] (ng/L)	Standard deviation at spike level (ng/L)	Number of values[f]	Number of non-qualified values[g]	MDL (ng/L)	"Valid" MDL?[h]	ML[i] (ng/L)	Detection level (ng/L)	Reporting level (ng/L)
4-Androstene-3,17-dione	Constant	3.0	5.2	1.1	0.20	8	1	**0.61**	Y	1.9	0.4	0.8
	Straight line	−0.7[k]	−4.3	2.2	0.77	8	0	2.3	N	--		
	Exponential	**0.42**	0.72	3.2	0.81	8	0	2.4	Y	7.8		
	Hybrid	**0.0**[k]	0.0	4.3	1.6	8	0	4.8	N	--		
Bisphenol A	Constant	4.3	7.5	1.1	0.43	8	0	1.3	N	--	10[j]	200
	Straight line	**1.6**	3.1	4.3	0.73	8	0	**2.2**	Y	7.0		
	Exponential	**1.8**	3.3	10.8	1.1	8	0	3.3	Y	10		
	Hybrid	**2.1**	3.6	21.7	1.1	8	0	3.2	N	--		
				32.5	1.9	8	0	5.8	N	--		
				43.4	2.0	8	0	6.1	N	--		
Cholesterol	Constant	57	98	0	19	8	0	57	N	--	102[j]	200
	Straight line	**43**	78	11	11	8	0	32	N	--		
	Exponential	**43**	78	43	16	8	0	48	N	--		
	Hybrid	**45**	79	108	14	8	0	**42**	Y	133		
				217	19	8	0	58	Y	184		
				325	18	8	0	53	N	--		
				434	29	8	0	86	N	--		
cis-Androsterone	Constant	**0.48**	0.83	0.4	0.12	7	2	**0.37**	Y	1.2	0.4	0.8
	Straight line	**0.36**	0.63	1.1	0.15	8	0	0.45	Y	1.4		
	Exponential	**0.36**	0.63	2.2	0.18	8	0	0.53	Y	1.7		
	Hybrid	**0.40**	0.69	3.2	0.19	8	0	0.56	N	--		
				4.3	0.19	8	0	0.57	N	--		
Dihydrotestosterone	Constant	2.5	4.3	1.1	0.46	8	4	1.4	N	--	2	4
	Straight line	1.8	3.6	2.2	0.99	8	3	3.0	N	--		
	Exponential	1.6	3.5	3.2	0.70	8	1	**2.1**	Y	6.7		
	Hybrid	**2.1**	3.9	4.3	0.79	8	0	2.4	Y	7.6		

Table 25. Estimates of Currie's critical level (Lc), the *intra*laboratory detection estimate (IDE), and the U.S. Environmental Protection Agency (USEPA) method detection limit (MDL) and minimum level (ML) concentrations for method analytes using ASTM standards D6091–07 (ASTM International, 2007) and D7510–10 (ASTM International, 2010).—Continued

[N, no; ng/L, nanograms per liter; Y, yes; --, not applicable; values in **bold** considered to exhibit good agreement for Lc by different models and with the estimated MDL]

Method analyte	Lc and IDE summary[a]			USEPA MDL and minimum level summary[a]							Detection and reporting levels applied in this report[b]	
	Standard deviation model[c]	Lc (ng/L)	IDE[d] (ng/L)	Spike level[e] (ng/L)	Standard deviation at spike level (ng/L)	Number of values[f]	Number of non-qualified values[g]	MDL (ng/L)	"Valid" MDL?[h]	ML[i] (ng/L)	Detection level (ng/L)	Reporting level (ng/L)
Epitestosterone	Constant	**0.99**	1.7	0.4	0.27	8	0	0.82	N	--	2	4
	Straight line	**0.79**	1.5	1.1	0.29	8	0	**0.85**	Y	2.7		
	Exponential	**0.79**	1.5	2.2	0.41	8	1	1.2	Y	3.9		
	Hybrid	**0.87**	1.5	3.2	0.34	8	0	1.0	Y	3.3		
				4.3	0.39	8	0	1.2	Y	3.8		
Equilenin	Constant	**1.3**	2.3	1.1	0.66	8	2[l]	2.0	N	--	1	2
	Straight line	**1.6**	2.6	2.2	0.48	8	0	**1.5**	Y	4.6		
	Exponential	**1.6**	2.6	3.2	0.55	8	0	1.6	Y	5.2		
	Hybrid	**1.3**	2.2	4.3	0.46	8	0	1.4	Y	4.4		
Equilin	Constant	**4.2**	7.2	1.1	1.4	8	6[m]	4.2	N	--	2	4
	Straight line	3.7	7.4	2.2	1.4	8	5[m]	4.3	N	--		
	Exponential	3.7	7.6	3.2	1.8	8	4	5.5	N	--		
	Hybrid	**3.9**	8.0	4.3	1.5	8	0	**4.6**	N	--		
Estriol	Constant	**1.1**	2.0	0.4	0.53	8	1[l]	1.6	N	--	1	2
	Straight line	**0.92**	1.7	1.1	0.18	8	0	0.55	Y	1.7		
	Exponential	**0.83**	1.5	2.2	0.37	8	0	1.1	Y	3.5		
	Hybrid	**0.93**	1.7	3.2	0.31	8	0	0.9	Y	3.0		
				4.3	0.55	8	0	1.7	Y	5.3		
Estrone	Constant	1.2	2.0	0.4	0.25	8	6	0.74	N	--	0.4	0.8
	Straight line	**0.64**	1.3	1.1	0.35	8	1	**1.0**	Y	3.3		
	Exponential	**0.67**	1.3	2.2	0.41	8	1	1.2	Y	4.0		
	Hybrid	**0.78**	1.4	3.2	0.65	8	0	1.9	Y	6.2		
				4.3	0.53	8	0	1.6	Y	5.0		
Mestranol	Constant	**0.96**	1.6	0.4	0.27	8	3	0.81	N	--	0.4	0.8
	Straight line	**0.76**	1.4	1.1	0.27	8	0	**0.82**	Y	2.6		
	Exponential	**0.77**	1.4	2.2	0.38	8	0	1.1	Y	3.6		
	Hybrid	**0.82**	1.4	3.2	0.29	8	0	0.88	Y	2.8		
				4.3	0.41	8	0	1.2	Y	3.9		
Norethindrone	Constant	1.2	2.1	0.4	0.40	8	0	1.2	N	--	0.4	0.8
	Straight line	**0.79**	1.6	1.1	0.28	8	0	**0.83**	Y	2.6		
	Exponential	**0.84**	1.6	2.2	0.44	8	0	1.3	Y	4.2		
	Hybrid	**0.90**	1.7	3.2	0.36	8	0	1.1	Y	3.4		
				4.3	0.66	8	0	2.0	Y	6.3		

Table 25. Estimates of Currie's critical level (Lc), the *intra*laboratory detection estimate (IDE), and the U.S. Environmental Protection Agency (USEPA) method detection limit (MDL) and minimum level (ML) concentrations for method analytes using ASTM standards D6091–07 (ASTM International, 2007) and D7510–10 (ASTM International, 2010).—Continued

[N, no; ng/L, nanograms per liter; Y, yes; --, not applicable; values in **bold** considered to exhibit good agreement for Lc by different models and with the estimated MDL]

Method analyte	Standard deviation model[c]	Lc (ng/L)[a]	IDE[d] (ng/L)	Spike level[e] (ng/L)	Standard deviation at spike level (ng/L)	Number of values[f]	Number of non-qualified values[g]	MDL (ng/L)	"Valid" MDL?[h]	ML[i] (ng/L)	Detection level (ng/L)	Reporting level (ng/L)
Progesterone	Constant	1.4	2.4	0.4	0.32	8	7	0.95	N	--	4	8
	Straight line	0.64	1.4	1.1	0.35	8	6	1.0	Y	3.3		
	Exponential	**0.78**	1.5	2.2	0.60	8	6	1.8	Y	5.8		
	Hybrid	**0.84**	1.6	3.2	0.28	8	0	**0.85**	Y	2.7		
				4.3	0.91	8	0	2.7	Y	8.7		
Testosterone	Constant	1.4	2.4	0.4	0.23	7	5	0.72	N	--	0.4	0.8
	Straight line	0.85	1.7	1.1	0.38	8	1	1.1	N	--		
	Exponential	0.81	1.6	2.2	0.54	8	0	1.6	Y	5.2		
	Hybrid	1.01	1.8	3.2	0.63	8	0	1.9	Y	6.0		
				4.3	0.45	8	0	1.4	Y	4.3		
trans-Diethylstilbestrol	Constant	0.51	0.87	0.4	0.16	8	0	0.49	N	--	0.4	0.8
	Straight line	**0.34**	0.62	1.1	0.15	8	0	**0.46**	Y	1.5		
	Exponential	**0.35**	0.62	2.2	0.08	8	0	0.23	N	--		
	Hybrid	**0.38**	0.66	3.2	0.23	8	0	0.69	Y	2.2		
				4.3	0.23	8	0	0.68	N	--		

[a]Summary of calculations provided by the D7501-10 DQCALC software using an earlier version that includes the exponential model. See "Analyte Detection Levels" section for additional information.

[b]Detection and reporting levels applied to validation data provided in this report and for data reporting in 2010 (and earlier for select analytes). Interim reporting level (IRL) type code was applied to all analytes, except 11-ketotestosterone, bisphenol A, cholesterol, and 3-*beta*-coprostanol, for which the minimum reporting level (MRL type code) convention was applied (Childress and others, 1999).

[c]The software calculates detection estimates for the constant, straight-line, exponential, and hybrid (Rocke-Lorenzato) models of the variation of intralaboratory standard deviation with concentration (ASTM International, 2007). The model selected as having the best fit parameters in DQCALC based on regression statistics and visual observation of plots is shown in italics.

[d]The IDE acronym can denote an interlaboratory detection estimate. but was applied, and is referred to here, as the *intra*laboratory detection estimate.

[e]For most analytes, there were no analyte detections at the 0-ng/L (unspiked) level and no or very few analyte detections at the 0.1-ng/L level; these levels were omitted from calculations. For several analytes, the 0.4-ng/L level also was omitted because of no or few detections.

[f]Number of determined values at the spiking level. One unusually high value at the 0.4-ng/L level was omitted for several analytes based on Grubbs' outlier testing.

[g]Number of determined values used in the calculations where the qualifying ion criteria were not met for the analyte (see "Analyte Detection Levels" section).

[h]MDL considered valid in DQCALC if the number of spike observations is 7 or more, and if the MDL does not exceed the spike concentration or is not less than 20 percent of the spike concentration.

[i]The USEPA minimum level is defined as 3.18 times the MDL and is only calculated by DQCALC when MDL is deemed valid.

[j]Analyte reported using MRL convention. Blank-based mean-offset MDL shown for comparison with other values.

[k]A substantial positive y-intercept can lead to less than zero (nonsensical) values.

[l]One of the non-qualified values was zero. no quantitation ion detected.

[m]Two of the non-qualified values were zero.

or less compared to the 4-ng/L applied interim detection level that was used because its quantitation ion is substantially more responsive than its secondary ions (at least six non-qualified values occurred even at the 2.2-ng/L spike level) and because of its more variable method performance. In general, the Lc and MDL values were similar to or somewhat greater than the Lc and MDL values determined before 2009 when using the original 13 IDSs (data not shown), indicating that GC/MS/MS instrumental sensitivity was slightly better in that earlier test for several analytes, a typical scenario for mass spectrometric (and other chromatographic-based) instrumentation.

The Lc and MDL values calculated for the blank-limited analytes BPA, cholesterol, and 3β-coprostanol were similar to or lower than one or more of the MDL and mean-off MDL values determined using the larger number of reagent-water blanks (table 22) and were at least four times lower than the applied MRL value. The Lc and MDL values for 11-ketotestosterone were greater than the mean-offset MDL from blanks (table 23) and similar to the applied MRL.

The detection data presented in tables 22, 23, and 25, coupled with other method performance observations including the previous detection study results (not presented in this report), were used to set the detection- and reporting-level values shown in table 10. These values were applied to sample data reported to NWIS by the filter-water (O–2434–12) and unfiltered-water (O–4434–12) methods beginning on October 1, 2011 (see section 12.2).

Holding-Time Experiments

Holding-time experiments that tested analyte stability in spiked reagent water stored refrigerated or frozen are summarized in this section. A test of storage stability of dry extracts before derivatization also is presented in this section because this storage condition is prescribed in section 9.14.4 and its use was required to complete the reagent-water holding-time experiments. The spiked-water tests were conducted using reagent water only; thus, the results might represent optimum stability compared to that obtainable with field matrices, especially matrices such as WWTP influent or primary effluent samples expected to have considerable microbiological activity. Except for the use of ascorbic acid as a dechlorination reagent (see section 8.4), no other sample preservation reagent currently (March 2012) is prescribed for use by this method. Schenck and others (2008) have shown that chlorination of water removes as much as 98 percent of tested steroid hormones. Ascorbic acid has been found to be an effective dechlorination reagent for many steroids and hormones (and other pharmaceuticals and personal-care products) determined in chlorinated-water test samples by USEPA method 1698 (U.S. Environmental Protection Agency, 2007a; 2010b), whereas sodium thiosulfate is the prescribed dechlorination reagent in USEPA method 539 (U.S. Environmental Protection Agency, 2010a) that determines

seven of the hormone analytes included in the method presented in this report.

Several studies by others have shown that some of the method analytes had rapid loss, or conversely formation, in tested environmental matrices stored refrigerated or even frozen, and that enhanced stability was obtained for some analytes by addition of certain preservative reagents including acids and biocides (Havens and others, 2010; U.S. Environmental Protection Agency, 2010b and references therein; Vanderford and others, 2011). USEPA method 539 prescribes addition of 2-mercaptopyridine-1-oxide sodium salt during sample collection to protect the seven hormone analytes determined by that method from microbial degradation. Limited testing of this preservation reagent using spiked reagent water samples resulted in no or very poor (<10 percent) recoveries for all eleven analytes that contain ketone functionality only (fig. 1), possibly because of reagent interference with the MSTFA derivatization step.

The procedure used for the holding-time experiments was based in part on the experimental and data-evaluation guidelines described in ASTM Standard Practice D4841–88 (ASTM International, 2008), although the number of day 0 replicates (four) was less than that prescribed for several analytes that have relative standard deviations greater than 9 percent. In each experiment, the analytes were fortified into the sample replicates at concentrations used to prepare the LRS samples and subsequently were treated as described in the sections that follow. The holding-time experiments were conducted before the IDS substitutions implemented in January 2010 (see section 10.7). Recovery data for all IDS compounds used in these experiments are reported in tables 26–28.

Analyte stability with time was evaluated by comparison of the mean concentration for quadruplicate spiked test samples held for various storage periods relative to the mean concentration from quadruplicate spikes analyzed at day 0. Three comparative approaches were applied to the results to gage overall stability. This was done to minimize misinterpretation of a compound's stability from any one comparative approach because recovery variation for day 0 replicates (to which other storage times are compared) was minimal for some analytes in these tests and leads to statistical predictions of instability that are not reflected by a substantial change in concentration for the storage period. The first comparison is that defined in ASTM Standard Practice D4841–88 where a tolerable range of variation (99-percent confidence interval) is calculated for each analyte using the standard error of the mean of determined concentrations from day 0 measurements (see details in ASTM International, 2008). The mean analyte concentration in replicates for a given storage period is then compared to this range. Those analytes with mean concentration for a storage period falling outside the tolerable range are considered to have undergone loss if below the range or to have undergone formation if above the range. The second, similar comparison was a Student's t-test of mean concentration for the storage period compared to the day 0 mean. For the *t*-test, a one-tailed (instead of a two-tailed) distribution, assuming either equal or unequal variance, was used for the comparisons as a more conservative evaluation because directionality (either toward only significant loss, especially, or toward only significant gain) in concentration for a given analyte was the important consideration. Probability (*p*) values of 0.01 or less were considered to indicate a significant difference in the means and, thus, a possible lack of stability for the storage period. The third comparison was of the value of the percentage change (PC) between the mean concentration for a given hold time (*Mean$_T$*) and the mean at day 0 (*Mean$_0$*) calculated as:

$$PC = 100 \times \frac{Mean_T - Mean_0}{Mean_0} \qquad (12)$$

A substantial negative PC value indicates loss during storage compared to day 0, whereas a substantial positive PC indicates analyte formation.

Accordingly, analyte instability was defined to occur for a storage period if the following three test criteria were met: (1) a mean concentration for the storage period that was outside the tolerable range, (2) a *p*-value of 0.01 or less from the *t*-test, and (3) a modulus (absolute value) PC greater than 20 percent; this 20 percent threshold also was used by the USEPA to indicate a substantial change in mean concentration in stability studies for hormones and other compounds (U.S. Environmental Protection Agency, 2010b). In tables 26–28, the PC values are shown in bold (non-italicized) if all three comparative criteria indicated instability. In some cases, indications of instability at shorter storage periods were not supported by data from longer storage periods, providing no clear evidence of compound instability. The mean (and standard deviation) of compound recovery relative to the amount spiked for the storage period also is shown in tables 26–28 for comparison to summarized method analyte and IDS recoveries from reagent-water spike samples provided in tables 12, 13, and 17, and IDS recoveries from reagent-water blank samples in table 18.

Stability of Method Analytes and Isotope-Dilution Standards When Stored Frozen as Dry Extracts

The holding-time experiment described in this section provides stability information for method analytes and IDS compounds for dry extracts contained in reaction vials that were held in a freezer (about −15°C) for various periods of time (section 9.14.4) just before compound derivatization and analysis. The dry-extract holding-time experiments were conducted alongside the refrigerator and freezer sample-storage experiments because extracts prepared during those hold-time studies also were stored for similar (or less) periods of time as dry extracts before derivatization and analysis.

The dry-extract test samples were prepared by adding 100 µL of the laboratory schedule 2434/4434 spike mixture containing the analytes (section 7.1.7) and 100 µL of the

IDS mixture (section 7.1.4; original IDSs used, see section 10.7) into 16 silanized reaction vials containing 2-mL dichloromethane. The solvent was evaporated to dryness in each reaction vial using nitrogen as described in section 9.14. Four of the 16 vials were immediately processed through the remainder of the method along with simultaneously prepared calibration standards (section 9.13) and analyzed by GC/MS/MS as day 0 dry-extract replicates.

The remaining 12 dry extracts were placed in a freezer at (−15±5°C). Four of these replicates were removed from the freezer after 8, 36, or 58 days of storage, which were holding periods that exceeded the maximum dry-extract storage period required to complete either the refrigerator or freezer holding tests. Upon completion of each storage period, the dry extracts were processed (along with corresponding groups of refrigerator or freezer test samples as described in the following subsections) through the remainder of the method as described for the day 0 dry-extract samples, and analyzed by GC/MS/MS using accompanying calibration standards that were prepared on the day that the dry-extract samples were removed from the freezer. Thus, the dry-extract replicates for each holding period were analyzed using independent instrumental calibrations.

A summary of the dry extract holding-time experiments is presented in table 26. All method analytes and IDS compounds appear to be stable (at least to loss) in dry extracts stored as long as 58 days, indicating that storage of sample extracts as dry, underivatized extracts in silanized reaction vials as described in section 9.14.4 provides good stability for all analytes for at least 58 days. Androstenedione and *cis*-androsterone had PC value just above 20 percent for day 58, yet their mean recoveries were no more than 95 percent (standard deviations <4 percent), and, thus, are not exhibiting actual "formation" on storage. The apparent increase in their concentration by day 58 was not thought to be an artifact of deuterium loss by the corresponding original IDS compounds whose recoveries also were >94% (see section 10.7).

Method Analyte Stability in Refrigerated Reagent Water

Analyte-only stability in reagent-water samples held refrigerated for 1, 3, and 8 days was tested to simulate possible storage periods for samples received at the NWQL and placed in a refrigerator only before sample extraction, and for samples maintained under refrigerated conditions (near 4°C) by field staff before and during shipment to the NWQL of sample coolers containing water ice.

Sixteen HDPE bottles containing about 450 mL of reagent water (section 6.6.1) were fortified with 100 µL of the laboratory schedule 2434/4434 spike mixture (section 7.1.7) containing the method analytes. No IDS compounds were added to the 12 bottles that were immediately placed in a refrigerator at 4±2°C. The remaining four day 0 replicates were fortified with 100 µL of the IDS mixture (section 7.1.4;

Table 26.   Summary of holding-time experiment for quadruplicate dry extracts containing method analytes and isotope-dilution standard compounds stored frozen (–15 degrees Celsius) from 0 to 58 days[a].

[SD, standard deviation of recovery; all values in percent; value in **bold** indicates a compound whose mean concentration for the storage period was classified as different from day 0 by all three applied test criteria[b]; value in **_bold italics_** indicates that the compound was outside the tolerable range and had a t-test p-value of 0.01 or less, but did not exceed the percentage change criteria[b]; value in _italics_ indicates that the compound had a t-test p-value of 0.01 or less but did not meet the other two criteria for classification as being different from day 0[b]]

Analyte	Day 0		Day 8			Day 36			Day 58		
	Mean recovery	SD	Mean recovery	SD	Change[c]	Mean recovery	SD	Change	Mean recovery	SD	Change
					Method analytes						
11-Ketotestosterone	83.4	3.8	76.5	5.4	–8.3	77.1	1.3	–7.5	92.0	1.8	_10.3_
17-_alpha_-Estradiol	86.7	2.5	88.3	2.8	1.9	87.6	1.5	1.0	89.6	2.2	3.3
17-_alpha_-Ethynylestradiol	88.7	3.9	89.7	0.6	1.2	86.2	1.5	–2.7	94.3	1.1	6.3
17-_beta_-Estradiol	95.6	2.2	94.8	3.9	–0.9	93.5	2.5	–2.1	92.7	2.3	–3.0
3-_beta_-Coprostanol	90.3	1.7	82.3	2.7	–8.9	79.7	1.3	_**–11.8**_	96.1	2.0	_**6.4**_
Androstenedione	78.8	4.7	88.7	2.2	12.5	87.9	2.5	_11.6_	94.9	3.8	**20.5**
Bisphenol A	87.1	1.5	76.0	2.4	–12.7	84.8	0.3	–2.6	93.2	2.9	_**7.1**_
Cholesterol	94.4	1.3	95.4	1.7	1.1	89.0	1.2	_**–5.7**_	99.9	2.6	_**5.9**_
cis-Androsterone	79.6	4.8	77.1	8.7	–3.1	81.5	3.9	2.4	95.8	3.6	**20.4**
Dihydrotestosterone	96.2	10.2	83.1	11.7	–13.6	99.6	11.9	3.5	105	4.5	9.0
Epitestosterone	85.9	1.5	76.1	1.8	–11.3	78.2	2.6	_**–8.9**_	89.9	1.7	_4.7_
Equilenin	79.2	3.6	78.3	5.3	–1.2	74.1	0.6	–6.4	91.4	3.1	_**15.4**_
Equilin	76.1	5.9	68.6	4.6	–9.8	85.4	5.1	12.3	87.8	4.5	_15.4_
Estriol	98.1	9.6	111	6.1	13.6	101	4.8	3.2	110	2.9	12.5
Estrone	98.0	1.5	103	5.8	4.8[d]	103	3.3	5.3[d]	97.9	1.7	–0.1
Mestranol	100	1.7	93.2	2.8	–7.1	91.2	2.8	_**–9.0**_	96.0	1.4	_–4.3_
Norethindrone	88.7	2.7	87.8	5.9	–1.1	86.4	4.4	–2.6	94.9	3.0	7.0
Progesterone	85.8	5.4	82.6	3.1	–3.7	82.1	1.2	–4.3	96.0	1.5	_11.9_
Testosterone	84.8	3.3	89.1	4.2	5.1	81.2	0.8	–4.3	87.6	1.5	3.3
trans-Diethylstilbestrol	84.4	1.8	86.0	3.8	1.8	84.5	4.3	0.1	91.3	2.4	_**8.1**_
					Isotope-dilution standards						
17-_beta_-Estradiol-d_4	90.7	2.7	95.4	1.5	5.2	90.9	1.8	0.2	97.2	3.7	7.2
Androstenedione-d_7	91.7	4.6	83.4	6.5	–9.1	85.0	3.2	–7.3	94.2	3.5	2.7
Bisphenol A-d_{16}	95.4	2.6	94.4	3.1	–1.0	105	0.9	_**10.5**_	94.9	3.5	–0.5
Cholesterol-d_7	101	2.6	97.7	5.9	–3.0	102	1.9	1.2	101	2.2	0.4
Diethylstilbestrol-d_8	93.1	2.6	90.9	2.3	–2.4	95.4	1.9	2.5	94.0	4.6	1.0
Dihydrotestosterone-d_4	98.0	4.8	94.7	11.7	–3.3	87.5	3.0	_–10.7_	96.9	4.7	–1.0
Estriol-d_3	91.6	8.5	80.8	7.0	–11.7	74.8	1.2	_–18.3_	86.8	3.6	–5.2
Estrone-d_4	90.1	3.0	85.2	2.3	–5.4	80.2	2.0	_**–10.9**_	94.4	3.8	4.8
Ethynylestradiol-d_4	92.3	5.6	94.2	4.6	2.0	87.1	3.6	–5.5	97.6	5.7	5.8
Mestranol-d_4	90.2	3.6	95.7	5.1	6.1	96.6	2.5	7.1	98.2	3.2	_8.9_
Norethindrone-d_6	92.0	4.9	83.0	2.8	–9.8	84.2	4.2	–8.5	96.2	2.7	4.5
Progesterone-d_9	87.5	4.0	84.9	5.0	–2.9	82.9	3.2	–5.2	94.9	3.4	8.5
Testosterone-d_5	90.2	2.7	91.0	5.8	0.8	88.0	3.7	–2.5	94.9	3.0	5.2

[a] Holding-time experiment preformed with original 13 isotope-dilution standards, 6 of which were subsequently removed from the method because of deuterium-loss issues.

[b] The mean concentration for the storage period is defined as being different from the mean concentration at day 0 by meeting all of the following criteria: (1) being outside the tolerable range (99-percent confidence interval) as described in D4841–88 (ASTM International, 2008), (2) having a p-value of 0.01 or less based on t-test (equal or unequal variance), and (3) having a modulus percentage change[c] at day T relative to day 0 greater than 20 percent (see "Holding-Time Experiments" section).

[c] The percentage change = 100 × (mean concentration at day T – mean concentration at day 0)/mean concentration at day 0.

[d] The mean concentration was outside the tolerable range only[b].

Table 27. Summary of holding-time experiment for quadruplicate reagent-water samples spiked with method analytes only and stored refrigerated (4 degrees Celsius) from 0 to 8 days[a].

[SD, standard deviation of recovery; all values in percent; --, not applicable; value in **bold** indicates a compound whose mean concentration for the storage period was classified as different from day 0 by all three applied test criteria[b]; value in ***bold italics*** indicates that the compound was outside the tolerable range and had a t-test p-value of 0.01 or less, but did not exceed the percentage change criteria[b]; value in *italics* indicates that the compound had a t-test p-value of 0.01 or less but did not meet the other two criteria for classification as being different from day 0[b]]

Analyte	Day 0		Day 1			Day 3			Day 8		
	Mean recovery	SD	Mean recovery	SD	Change[c]	Mean recovery	SD	Change	Mean recovery	SD	Change
Method analytes											
11-Ketotestosterone	79.6	2.7	82.2	4.6	3.2	84.3	6.4	4.6	79.8	6.3	–1.1
17-*alpha*-Estradiol	92.2	2.0	92.8	3.7	0.7	97.7	4.8	4.7	95.0	2.4	1.7
17-*alpha*-Ethynylestradiol	84.9	5.5	89.8	4.0	5.8	83.4	1.5	–3.1	86.1	3.2	0.1
17-*beta*-Estradiol	89.8	3.4	92.1	2.3	2.5	96.2	2.9	5.8	95.1	2.1	4.5
3-*beta*-Coprostanol	85.1	2.2	84.2	6.1	–1.2	79.3	6.9	–8.1	82.0	13.6	–5.1
Androstenedione	89.4	2.5	92.7	5.6	3.7	102	7.9	12.6[d]	92.7	1.8	2.4
Bisphenol A	72.8	0.7	74.5	2.0	2.3	73.7	1.1	–0.1	77.7	2.7	*5.4*
Cholesterol	89.4	3.4	87.7	5.1	–1.9	82.0	6.4	–9.4	88.8	11.5	–2.0
cis-Androsterone	84.2	18.6	71.7	10.9	–15.0	63.5	11.3	–25.5	63.5	11.4	–25.6
Dihydrotestosterone	75.2	10.3	75.5	12.0	0.3	63.8	6.3	–16.2	73.4	9.2	–3.7
Epitestosterone	83.3	7.4	80.5	7.6	–3.6	87.0	8.6	3.0	74.9	2.6	–11.3
Equilenin	67.7	2.4	63.0	3.6	–6.9	68.4	9.2	–0.1	69.9	7.0	2.0
Equilin	64.2	3.4	70.7	8.1	10.0	72.7	9.0	11.8	57.5	4.7	–11.6
Estriol	102	5.6	110	7.3	7.3	105	9.3	1.9	112	7.5	8.3
Estrone	99.5	4.9	106	15.7	6.3	124	10.8	**22.9**	102	3.6	1.4
Mestranol	87.1	4.5	61.8	4.9	**–29.0**	55.1	6.2	**37.5**	53.0	4.1	**–40.0**
Norethindrone	83.1	3.1	83.7	5.9	0.7	85.2	8.1	1.3	84.5	1.0	0.3
Progesterone	74.5	3.8	61.9	16.2	–17.1	60.1	13.6	–20.2[d]	43.7	3.2	**–42.2**
Testosterone	85.7	2.8	90.3	6.4	5.4	94.0	8.4	8.3	87.3	3.2	0.6
trans-Diethylstilbestrol	82.2	2.2	79.8	2.8	–2.9	73.7	0.8	*–11.5*	68.8	6.6	*–17.4*
Isotope-dilution standards											
17-*beta*-Estradiol-d_4	68.3	4.2	67.9	2.3	--	61.6	2.5	--	61.9	3.2	--
Androstenedione-d_7	60.0	3.5	55.7	5.3	--	44.7	7.0	--	53.1	3.9	--
Bisphenol A-d_{16}	76.5	5.1	73.2	2.5	--	69.0	4.0	--	68.2	2.9	--
Cholesterol-d_7	65.2	2.0	63.2	4.0	--	63.0	2.6	--	60.9	2.8	--
Diethylstilbestrol-d_8	64.4	3.4	62.3	2.7	--	58.8	4.2	--	53.8	9.4	--
Dihydrotestosterone-d_4	65.8	15.6	66.2	9.9	--	70.2	8.6	--	65.0	10.2	--
Estriol-d_3	82.9	4.1	77.8	3.6	--	83.0	14.2	--	78.3	6.4	--
Estrone-d_4	64.9	7.0	58.3	8.0	--	47.2	5.2	--	59.1	1.6	--
Ethynylestradiol-d_4	66.0	7.7	63.3	2.9	--	59.7	2.0	--	59.6	2.1	--
Mestranol-d_4	62.4	4.7	64.4	2.5	--	59.3	2.3	--	59.1	1.3	--
Norethindrone-d_6	61.7	2.7	58.7	3.6	--	49.1	5.1	--	55.9	3.6	--
Progesterone-d_9	51.4	4.7	48.3	8.8	--	36.5	4.2	--	44.9	3.0	--
Testosterone-d_5	65.4	4.8	62.6	5.0	--	53.6	6.0	--	59.8	2.5	--

[a]Holding-time experiment preformed with original 13 isotope-dilution standards, 6 of which were subsequently removed from the method because of deuterium-loss issues.

[b]The mean concentration for the storage period is defined as being different from the mean concentration at day 0 by meeting all of the following criteria: (1) being outside the tolerable range (99-percent confidence interval) as described in D4841–88 (ASTM International, 2008), (2) having a p-value of 0.01 or less based on t-test (equal or unequal variance), and (3) having a modulus percentage change[c] at day T relative to day 0 greater than 20 percent (see "Holding-Time Experiments" section).

[c]The percentage change = 100 × (mean concentration at day T – mean concentration at day 0)/mean concentration at day 0.

[d]The mean concentration was outside the tolerable range[b].

Table 28. Summary of holding-time experiment for quadruplicate reagent-water samples spiked with method analytes only and stored frozen (−15 degrees Celsius) from 0 to 56 days[a].

[SD, standard deviation of recovery; all values in percent; --, not applicable; value in **bold** indicates a compound whose mean concentration for the storage period was classified as different from day 0 by all three applied test criteria[b]; value in ***bold italics*** indicates that the compound was outside the tolerable range and had a t-test p-value of 0.01 or less, but did not exceed the percentage change criteria[b]; value in *italics* indicates that the compound had a t-test p-value of 0.01 or less but did not meet the other two criteria for classification as being different from day 0[b]]

Analyte	Day 0 Mean recovery	Day 0 SD	Day 2 Mean recovery	Day 2 SD	Day 2 Change[c]	Day 7 Mean recovery	Day 7 SD	Day 7 Change	Day 14 Mean recovery	Day 14 SD	Day 14 Change	Day 21 Mean recovery	Day 21 SD	Day 21 Change	Day 56 Mean recovery	Day 56 SD	Day 56 Change
Method analytes																	
11-Ketotestosterone	81.7	2.5	83.7	7.3	1.4	80.8	1.4	-2.4	83.7	2.5	1.2	87.0	3.5	4.5	90.6	4.3	8.9
17-*alpha*-Estradiol	93.0	0.8	90.5	1.3	-3.8	93.3	1.3	-1.0	93.0	3.8	-1.4	93.8	1.7	-1.1	91.5	1.0	-3.4[d]
17-*alpha*-Ethynylestradiol	89.2	4.1	88.7	2.4	-1.7	86.0	2.6	-4.9	87.2	2.8	-3.6	86.3	2.3	-5.1	93.2	3.0	2.7
17-*beta*-Estradiol	94.7	1.7	98.0	3.3	2.4[d]	95.6	1.2	-0.3	97.4	3.8	1.4	99.8	1.9	3.5	89.0	1.7	***-7.6***
3-*beta*-Coprostanol	79.4	2.9	58.2	5.0	-27.5	54.4	1.8	**-32.4**	56.4	5.5	**-30.0**	54.9	1.8	**-32.1**	68.8	7.4	-14.9[d]
Androstenedione	92.0	8.5	103	9.8	11.0	92.6	4.1	-0.7	106	9.8	13.3	110	20.3	17.2	95.6	9.3	2.1
Bisphenol A	83.8	1.6	84.6	2.8	-0.2	83.6	1.7	-1.6	87.8	1.2	3.3	87.0	1.4	1.8	99.2	4.0	***16.2***
Cholesterol	82.1	2.0	62.9	4.7	-24.3	58.6	2.6	**-29.5**	62.1	5.6	**-25.4**	60.6	3.3	**-27.6**	71.9	7.8	-13.9[d]
cis-Androsterone	77.9	6.1	64.8	5.0	-17.7	77.3	2.0	-1.9	68.3	10.6	-13.6	69.3	10.1	-12.7	77.5	6.7	-2.2
Dihydrotestosterone	84.8	9.1	75.5	2.9	-11.9	85.1	7.6	-0.9	74.9	4.5	-12.8	80.4	7.5	-6.9	92.7	4.5	7.3
Epitestosterone	82.1	6.2	81.6	3.6	-1.7	81.4	3.5	-2.2	83.8	8.3	0.6	83.4	7.7	-0.3	87.8	7.3	5.1
Equilenin	67.4	7.0	57.7	4.6	-15.3	58.5	8.3	-14.3	61.0	0.9	-10.7	64.0	4.5	-6.8	83.2	3.1	21.3
Equilin	87.1	18.8	89.5	8.0	1.6	80.0	15.0	-9.4	106	16.2	20.3	111	18.4	24.8	113	18.2	27.6
Estriol	98.0	3.0	95.7	5.7	-3.4	102	6.9	2.6	99.6	2.6	0.3	97.4	4.7	-2.4	107	3.2	7.0
Estrone	115	26.3	125	8.4	8.2	103	2.3	-11.4	119	10.0	2.2	127	17.0	8.8	102	9.2	-12.8
Mestranol	79.7	3.0	60.4	3.3	**-25.1**	65.9	6.4	***-18.4***	62.2	6.8	**-23.0**	60.5	4.4	**-25.4**	65.2	4.9	***-19.7***
Norethindrone	87.6	5.7	87.2	4.8	-1.6	86.8	3.4	-2.2	91.4	4.4	2.8	91.7	8.4	2.6	90.3	2.1	1.2
Progesterone	79.2	17.0	65.8	9.7	-17.8	57.1	9.9	-28.9	64.9	8.8	-19.2	69.5	20.0	-13.8	67.5	14.3	-16.2
Testosterone	83.4	4.4	88.3	3.8	4.7	84.1	2.8	-0.5	87.8	4.0	3.7	91.8	7.2	8.0	89.3	4.8	5.1
trans-Diethylstilbestrol	84.8	2.5	80.2	1.7	-6.5	81.8	0.4	-4.8	82.3	1.9	-4.2	81.0	2.0	-6.3	83.0	3.7	-3.9
Isotope-dilutions standards																	
17-*beta*-Estradiol-d_4	77.3	2.9	81.3	2.4	--	74.3	4.5	--	61.9	3.2	--	61.9	3.2	--	82.6	4.2	--
Androstenedione-d_7	70.4	8.5	61.0	3.8	--	73.2	2.7	--	53.1	3.9	--	53.1	3.9	--	74.4	7.4	--
Bisphenol A-d_{16}	98.6	3.5	94.9	5.6	--	90.7	9.5	--	68.2	2.9	--	68.2	2.9	--	90.5	2.5	--
Cholesterol-d_7	69.6	3.6	69.9	7.4	--	70.6	2.2	--	60.9	2.8	--	60.9	2.8	--	72.4	2.7	--
Diethylstilbestrol-d_8	75.6	3.6	72.3	4.5	--	65.9	11.7	--	53.8	9.4	--	53.8	9.4	--	78.1	2.1	--
Dihydrotestosterone-d_4	80.2	5.5	86.5	6.7	--	79.1	8.0	--	65.0	10.2	--	65.0	10.2	--	86.6	4.1	--
Estriol-d_3	64.3	4.0	69.0	5.7	--	59.9	6.1	--	78.3	6.4	--	78.3	6.4	--	69.2	3.1	--
Estrone-d_4	63.6	13.9	56.6	5.7	--	66.2	5.0	--	59.1	1.6	--	59.1	1.6	--	74.9	8.9	--
Ethynylestradiol-d_4	72.0	5.5	74.7	2.7	--	69.4	6.2	--	59.6	2.1	--	59.6	2.1	--	78.8	3.9	--
Mestranol-d_4	78.9	4.1	79.5	1.6	--	79.4	3.4	--	59.1	1.3	--	59.1	1.3	--	80.2	3.2	--
Norethindrone-d_6	70.0	7.6	66.5	2.4	--	73.4	3.3	--	55.9	3.6	--	55.9	3.6	--	76.5	4.5	--
Progesterone-d_9	55.3	14.0	42.7	7.3	--	54.3	2.8	--	44.9	3.0	--	44.9	3.0	--	59.5	6.3	--
Testosterone-d_5	73.5	5.7	70.5	2.6	--	75.5	2.6	--	59.8	2.5	--	59.8	2.5	--	78.3	3.7	--

[a] Holding-time experiment preformed with original 13 isotope-dilution standards, 6 of which were subsequently removed from the method because of deuterium-loss issues.

[b] The mean concentration for the storage period is defined as being different from the mean concentration at day 0 by meeting all of the following criteria: (1) being outside the tolerable range (99-percent confidence interval) as described in D4841–88 (ASTM International, 2008), (2) having a p-value of 0.01 or less based on t-test (equal or unequal variance), and (3) having a modulus percentage change[c] at day T relative to day 0 greater than 20 percent (see "Holding-Time Experiments" section).

[c] The percentage change = 100 × (mean concentration at day T − mean concentration at day 0)/mean concentration at day 0.

[d] The mean concentration was outside the tolerable range only.

original IDSs used, see section 10.7), immediately extracted, further processed through step 9.14.4 of the method, and stored in a freezer as dry extracts in silanized reaction vials for subsequent processing with the refrigerated hold-time sample extracts as described as follows.

Four test-sample replicates were removed from the refrigerator after 1, 3, or 8 days of storage, and immediately fortified with the IDS compounds (100 µL of the IDS mixture), extracted, and further processed through step 9.14.4 (the dry-extract stage). Day 1 and day 3 extracts were held frozen with the day 0 extracts (and the day 8 dry-extract test samples; see the immediate previous subsection) until the day 8 refrigerated samples were processed through step 9.14.4. All test sample extracts, along with accompanying calibration standards, were processed through the remainder of the method and analyzed by GC/MS/MS.

A summary of the refrigerated sample holding-time experiments is presented in table 27. Only two analytes had clearly significant concentration changes in refrigerated reagent water through 8 days of storage. Mestranol showed the largest initial drop in concentration (PC of –29 percent) at day 1, but the loss rate appeared to rapidly diminish with further storage because its PC values at day 3 and day 8 were no more than –40 percent. Progesterone also appreared to undergo significant loss with a PC by day 8 of –42 percent. The analyte *cis*-androsterone had PC values <–26 percent at days 3 and 8, but the changes were not statistically significant because of the relatively large variability (19 percent standard deviation) at day 0. Similar holding-time experiments by the U.S. Environmental Protection Agency (2010b, pages 17–19 and 38) found that *cis*-androsterone's concentration decreased (mean PC of about –42 percent on day 7) in "chlorinated effluent samples, held in HDPE bottles, dechlorinated with ascorbic acid, and stored at 4°C (for as much as 14 days), with no pH adjustment." Interestingly, that study reported a statistically significant loss (PC of –30 percent) for mestranol in day 7 test samples, but no apparent loss in day 14 samples compared to day 0 concentrations. That study also showed (page 38) a non-statistically significant decrease in progesterone concentration (mean PC of about –37 percent by day 14).

Method Analyte Stability in Frozen Reagent Water

Analyte-only stability in reagent-water samples held frozen (–15±5°C) for 2, 7, 14, 21, and 56 days was tested to simulate some potential frozen-sample storage periods. Use of HDPE bottles facilitates freeze storage of water samples in comparison to use of glass bottles that are susceptible to breakage or Teflon® bottles that are more susceptible to seam failure or cap leakage when frozen. Also, freezer storage was believed preferred for maintaining analyte concentrations or at least slowing analyte-loss processes relative to refrigerated conditions, especially for environmental samples. Following

sample login at the NWQL, samples that are not expected to be extracted within 3 days are placed in a freezer for storage until they can be extracted. Note: Upon login, samples are automatically placed in a freezer for storage unless indentified by method staff to be placed in a refrigerator.

Twenty-four HDPE bottles containing about 450 mL of reagent water were fortified with 100 µL of the laboratory schedule 2434/4434 spike mixture (section 7.1.7) containing the method analytes. No IDS compounds were added to the 20 bottles that were immediately placed in a freezer. The remaining four day 0 replicates were fortified with 100 µL of the IDS mixture (section 7.1.4; the original IDSs were used, see section 10.7), immediately extracted, processed through step 9.14.4, and stored frozen as dry extracts for subsequent processing with the day 2–21 freezer hold-time sample extracts as described as follows.

Four test-sample replicates were removed from the freezer after 2, 7, 14, or 21 days of storage. The frozen replicates were allowed to thaw at room temperature for about 18 h, and then immediately fortified with the IDS compounds (100 µL of the IDS mixture), extracted, and further processed through step 9.14.4 (to dry-extract stage). The resultant day 2, 7, 14, and 21 extracts were held frozen with the day 0 extracts and the day 36 dry-extract test samples (see the "Stability of Method Analytes and Isotope-Dilution Standards When Stored Frozen as Dry Extracts" section). These were processed 36 days after initially spiking, along with accompanying calibration standards, through the remainder of the method and analyzed by GC/MS/MS.

The remaining four samples were removed from the freezer after 56 days, thawed, and immediately processed through the entire method. Derivatization was completed along with the day 58 dry-extract test samples and accompanying calibration standards before GC/MS/MS analysis.

A summary of the frozen sample holding-time experiment is shown in table 28. Seventeen analytes, including *cis*-androsterone, had no statistically significant concentration changes in frozen reagent water through 56 days of storage. Mestranol concentrations initially dropped after 2 days of freezer storage (PC of –25 percent), but exhibited no further decrease for all other storage periods. The amount of mestranol loss for all frozen-sample storage periods is less than the loss observed after only one day of refrigerated storage (table 27), which suggests that most of the mestranol loss occurs when the water sample is warmer (for example during the initial sample cool-down or thaw periods) rather than at colder (freezer) temperatures. Cholesterol and 3β-coprostanol had significant decreases in concentration (PC ≤–32 percent for both) in all but the longest (day 56) frozen samples that might be related to sorption issues for these two sterols because of the decrease in their water solubilities with decreasing temperature (see previous description of sterol solubility/sorption issues in the "Cholesterol-d_7 Recoveries in Non-salted Reagent Water" section).

Progesterone concentrations were not significantly different for any storage period, although the PC at day 7 (–29 percent) was substantial. Equilin concentrations appeared to increase (PC ≤28 percent) in day 14–56 samples, although the mean equilin recoveries for these storage periods were no more than 113 percent and the changes were not significant. As described in the Surface Water section for progesterone and the Secondary Wastewater Effluent section for equilin, substantially greater variability in determined concentrations by this method reduces the ability to detect statistically significant trends under all storage conditions for at least these two analytes.

These experiments indicate that freeze-storage of samples for at least 56 days does not significantly alter sample concentrations for most method analytes, especially relative to storage in a refrigerator. The possible exceptions are cholesterol and 3β-coprostanol because of presumed sorption losses, and mestranol because of rapid initial loss (day 1), the rate of which slows with additional freezer storage. Sample freezing is anticipated to reduce biotic activity in sample matrices relative to refrigerated storage conditions. Thus, storage of samples in a freezer is prescribed as the standard storage condition for all samples, unless they can be extracted within 3 days of receipt at the NWQL. Freezer storage of samples also is prescribed for field samples that can not be shipped immediately from the field on water ice.

Compound Recoveries in Other Spiked-Matrix Samples

Recoveries of method analytes from various field matrices that were spiked just before extraction (exceptions noted in this section) with the analytes and IDS compounds at the same fortification concentrations as used for the LRSs are shown in table 29. These samples are grouped by sample medium (matrix) and secondarily by being a filtered (LS 2434) or unfiltered (LS 4434) sample. The spiked-matrix samples include field-requested laboratory matrix-spike samples (FRLMSs) and laboratory matrix-spike samples (MSPKs) (see section 13.4). Although some matrices are WWTP effluent samples, none contained ascorbic acid because these samples were collected from WWTPs that do not use chlorination disinfection. Although the recovery calculation includes ambient analyte concentration correction, analyte recovery values shown in table 29 for those spiked matrices that had corresponding ambient-sample concentrations from 10 to 100 percent (value in bold) or from 101 to 225 percent (value in bold italics) of the fortified amount are highlighted to indicate potential ambient bias. Bias is clearly evident for cholesterol and 3β-coprostanol in the FRLMS surface-water sample collected on October 19, 2010.

Also shown in table 29 are a groundwater sample (labeled FRLMS 83) and a treated water-supply sample (FRLMS 85; does contain ascorbic acid) collected in South Dakota that were spiked upon receipt with the method analytes only

(no IDS compounds) and stored frozen for 83 and 85 days, respectively. Two reagent-water samples (RWS 83 and RWS 85) were simultaneously spiked and stored for those periods; neither sample included ascorbic acid. Except for equilenin and *trans*-diethylstilbestrol in FRLMS 83 (as described below in this section), the recoveries of all analytes in these stored samples were within the recovery range for the other spiked matrices. The statistical summary shown in table 29 includes all spiked field matrices including FRLMS 83 and 85, but not the reagent-water spikes.

Comparisons of analyte method recoveries relative to IDS absolute recoveries for the FRLMS and MSPK samples are shown in figure 10 and omit analytes with ambient concentrations exceeding 100 percent of the fortification concentration.

Mean analyte recoveries in the matrix-spike samples (table 29) ranged from 66 percent (progesterone) to 141 percent (3β-coprostanol). RSDs were ≤25 percent for all but six analytes. High ambient concentrations disproportionally bias some of these means, especially for cholesterol and 3β-coprostanol. Median recoveries range from 70 percent (progesterone) to 105 percent (*cis*-androsterone and epitestosterone). For most analytes, mean and median recoveries differed by no more than 5 percent. The relative F-pseudosigma values were less than 22 percent for all analytes except equilin (36 percent RF_o), which had a broad recovery range (41–173 percent), and progesterone (65 percent RF_o), which had an even broader recovery range (0–271 percent). As previously noted, sample concentrations for equilin and progesterone are reported to NWIS as estimated. Two non-detections and all other low (<46 percent) progesterone recoveries occurred in the unfiltered surface-water matrices. As noted in the "Surface Water" section that describes validation-matrix results for the Rapid Creek surface-water matrix, the cause of these progesterone losses is not known. The recovery of progesterone's IDS, medroxyprogesterone-d_3, was relatively low in some of these samples, but not in others. Loss of progesterone in these matrices clearly was not well emulated by this non-exact IDS analog. Replacement of medroxyprogesterone-d_3 with exact IDS analog progesterone-$^{13}C_3$ on March 1, 2012, is likely to improve progesterone-to-IDS emulation (see section 10.7) in all matrices based on the performance data obtained for other analytes that have exact IDS analogs as shown in this report.

Estriol and 11-ketotestosterone had unusually low recoveries (<31 percent) in the same four unfiltered samples (the three 6/29 NY and one 7/13 NY samples), comprising three different matrix types. Recoveries of their corresponding IDSs (table 7) likewise clearly did not emulate the analytes' absolute recoveries in these matrices because the method analyte recoveries were lower than the IDS compound recoveries. Possible reasons for these low recoveries were described in the "Analyte Method Recoveries in Laboratory Reagent-Water Spike Samples" section.

Recoveries for most method analytes in the two matrices (FRLMS 83 and 85) that were stored frozen for as

Table 29. Recoveries of method compounds in field-requested laboratory matrix-spike samples (FRLMS) and laboratory matrix-spike samples (MSPK) submitted in 2010 for analysis by laboratory schedules 2434 or 4434.

[RSD, relative standard deviation. All values in percent except as otherwise noted. Some values might have additional bias due to concentrations in the ambient sample between 10 and 100 percent (**bold** values) or 100 and 225 percent (***bold italicized*** values) of the fortified amount. Analyte fortification levels were 10 nanograms per liter (ng/L) for 17 analytes, 100 ng/L for bisphenol A, 320 or 1,000 ng/L for 3-*beta*-coprostanol, and 1,000 ng/L for cholesterol assuming a 0.5-L sample volume]

Spiked sample type:	MSPK	MSPK	MSPK	MSPK	FRLMS	MSPK	FRLMS	FRLMS	FRLMS	FRLMS	FRLMS	FRLMS	FRLMS	FRLMS	FRLMS
Method:	4434	2434	2434	2434	4434	4434	4434	4434	4434	4434	4434	4434	4434	4434	4434
Collected (month/day/2010):	7/13	5/25	5/25	5/26	7/29e	6/29	9/22	9/23	9/28	9/29	9/30	10/6	10/6	10/13	10/19
State:	N.Y.	Minn.	Minn.	Minn.	S. Dak.	N.Y.	N.Y.	Minn.	Minn.	Minn.	Minn.	Ohio	Ohio	Wis.	Mich.
Sample medium:	Blended, untreated water supply	Ground-water	Ground-water	Ground-water	Ground-water	Surface water	Surface water	Surface water	Surface water	Surface water	Surface water	Surface water	Surface water	Surface water	Surface water
Method analytes															
11-Ketotestosterone	11.8	92.5	88.3	92.0	81.9	30.6	64.6	97.1	90.8	100	92.9	72.8	91.7	55.6	85.2
17-*alpha*-Estradiol	108	102	111	105	86.4	93.1	104.4	96.8	94.7	101	106	103	92.3	98.0	98.5
17-*alpha*-Ethynylestradiol	92.1	93.1	96.1	93.5	98.7	89.5	87.2	83.7	84.2	89.5	95.6	89.7	84.7	80.8	79.8
17-*beta*-Estradiol	102	97.4	104	96.1	98.7	85.0	97.6	97.6	94.2	99.3	98.8	89.0	90.3	87.9	87.9
3-*beta*-Coprostanol	92.1	91.4	95.5	98.0	88.0	**87.9**	**92.9**	**115**	**74.8**	94.0	97.2	86.5	89.9	89.7	***1,210***
4-Androstene-3,17-dione	98.9	88.8	97.7	95.7	92.4	108	113	99.6	109	108	110	115	94.7	117	130
Bisphenol A	94.2	86.6	87.3	**85.2**	87.9	80.4	**91.6**	**102**	**69.2**	81.7	89.1	80.0	84.7	71.1	75.7
Cholesterol	**92.6**	103	94.6	99.9	94.5	**89.3**	**85.1**	**99.9**	**77.6**	**90.1**	**100**	**84.1**	**84.5**	**97.6**	776
cis-Androsterone	85.2	122	116	101	126	114	135	105	107	106	112	121	92.2	160	128
Dihydrotestosterone	85.9	101	98.1	96.1	134	78.5	102	94.0	101	89.5	110	115	93.8	128	103
Epitestosterone	94.8	107	105	99.2	106	107	115	110	114	104	111	112	98.5	129	127
Equilenin	107	89.3	102	90.9	68.3	57.0	94.7	79.5	75.8	72.4	84.7	83.5	77.0	41.2	79.8
Equilin	88.2	80.1	71.4	97.2	75.5	68.1	103	145	129	134	127	90.6	131	89.8	41.2
Estriol	6.1	83.0	91.0	88.2	63.3	14.1	82.3	80.5	81.0	86.0	94.5	84.5	73.1	80.1	84.5
Estrone	103	101	99.9	99.8	104	90.4	102	**99.2**	**83.9**	95.4	102	97.3	97.3	93.5	110
Mestranol	101	95.2	101	97.6	99.9	86.9	97.4	96.8	87.5	97.9	103	91.0	90.1	90.3	86.6
Norethindrone	99.8	86.1	88.1	93.3	91.8	81.8	92.7	94.2	87.4	94.8	93.8	94.2	92.3	86.9	92.9
Progesterone	271	70.1	70.9	80.5	85.6	36.6	8.5	30.0	18.0	45.7	20.5	6.5	28.2	0.0	0.0
Testosterone	97.8	99.9	95.9	96.9	131	107	111	96.4	101	94.5	106	106	98.2	114	107
trans-Diethylstilbestrol	91.9	90.9	96.7	91.5	89.0	84.7	86.4	87.2	80.6	80.1	97.6	90.5	82.1	68.1	73.4
Isotope-dilution standards															
16-Epiestriol-d_2	42.4	75.1	65.3	82.0	65.1	60.5	74.6	65.9	63.5	54.3	80.9	69.8	91.4	66.9	65.7
17-*alpha*-Ethynylestradiol-d_4	92.6	70.0	73.2	83.2	51.5	80.4	77.3	67.3	71.1	64.2	87.8	69.6	90.6	67.2	70.1
17-*beta*-Estradiol-$^{13}C_6$	86.8	72.9	75.8	85.1	57.7	88.2	75.8	67.5	69.0	61.5	89.3	72.7	88.2	71.9	68.3
Bisphenol A-d_{16}	77.4	78.3	100	88.0	54.3	90.4	90.8	76.2	63.5	58.4	104	87.1	86.9	93.1	90.2
Cholesterol-d_7	85.1	68.3	56.0	70.2	66.6	92.2	77.0	78.2	85.1	83.5	75.3	79.4	77.0	75.0	64.2
Diethylstilbesterol-d_8	54.7	54.2	76.3	66.5	39.9	37.9	42.8	61.4	44.4	19.9	62.9	46.9	47.3	34.5	37.5
Estrone-$^{13}C_6$	89.1	71.4	81.1	88.0	34.1	85.7	74.6	71.2	73.0	66.7	93.2	76.9	89.7	71.4	68.8
Medroxyprogesterone-d_3	32.6	76.5	77.6	92.3	48.5	76.4	17.3	63.0	56.6	65.6	62.2	27.5	87.0	9.2	27.2
Mestranol-d_4	87.2	69.0	71.1	81.2	51.6	90.6	72.9	62.2	69.4	63.1	86.0	72.1	88.0	70.1	73.2
Nandrolone-d_3	119	60.9	74.5	88.5	30.3	85.0	55.3	65.0	63.0	66.9	81.7	55.8	86.6	41.6	44.5
Sample volume, milliliters	782	450	436	460	475	797	931	375	473	468	449	442	436	455	443

Table 29. Recoveries of method compounds in field-requested laboratory matrix-spike samples (FRLMS) and laboratory matrix-spike samples (MSPK) submitted in 2010 for analysis by laboratory schedules 2434 or 4434.—Continued

[RSD, relative standard deviation. All values in percent except as otherwise noted. Some values might have additional bias due to concentrations in the ambient sample between 10 and 100 percent (**bold** values) or 100 and 225 percent (***bold italicized*** values) of the fortified amount. Analyte fortification levels were 10 nanograms per liter (ng/L) for 17 analytes, 100 ng/L for bisphenol A, 320 or 1,000 ng/L for 3-*beta*-coprostanol, and 1,000 ng/L for cholesterol assuming a 0.5-L sample volume]

Spiked sample type: Method: Collected (month/day/2010): State: Sample medium:	FRLMS 2434 9/15 Wash. WWTP effluent	FRLMS 2434 9/15 Wash. WWTP effluent	MSPK 4434 6/29 N.Y. WWTP effluent	MSPK 4434 6/29 N.Y. WWTP effluent	MSPK 4434 6/30 N.Y. WWTP effluent	FRLMS 4434 9/23 Minn. WWTP effluent	FLRMS 83[a] 4434 2/17 S. Dak. Groundwater	FRLMS 85[a,b] 4434 2/17 S. Dak. Treated water supply	RWS 83[a] 4434 2/17 S. Dak. Reagent water	RWS 85[a] 4434 2/17 S. Dak. Reagent water
					Method analytes					
11-Ketotestosterone	98.4	92.0	12.2	11.5	123	113	99.4	91.3	98.1	87.5
17-*alpha*-Estradiol	109	104	108	110	88.6	95.3	101	110	111	103
17-*alpha*-Ethynylestradiol	99.2	98.1	99.3	95.5	90.9	88.7	98.1	100	104	96.2
17-*beta*-Estradiol	104	101	101	110	100	99.7	99.9	108	110	101
3-*beta*-Coprostanol	89.3	85.7	**97.3**	95.6	99.3	**98.6**	87.0	91.1	66.4	73.4
4-Androstene-3,17-dione	103	103	105	97.6	118	97.7	102	94.5	95.4	81.8
Bisphenol A	88.3	84.3	90.2	95.2	84.0	**74.2**	80.3	75.8	82.4	72.3
Cholesterol	90.3	86.2	**96.5**	92.4	97.1	***103***	65.2	76.8	52.4	65.1
cis-Androsterone	103	94.5	84.8	88.6	99.5	95.7	103	101	89.9	86.7
Dihydrotestosterone	93.0	93.9	86.5	85.7	126	106	94.5	95.2	90.9	80.2
Epitestosterone	103	96.0	96.6	101	114	105	103	100	101	87.0
Equilenin	84.1	73.3	103	98.7	68.4	86.2	49.2	95.5	97.9	84.8
Equilin	118	135	80.8	100	173	124	73.2	97.0	74.1	81.7
Estriol	105	96.3	9.6	16.1	49.4	75.5	99.4	102	103	96.3
Estrone	108	98.8	113	100	101	**92.1**	116	109	106	99.2
Mestranol	105	97.4	98.8	95.3	97.1	99.7	101	92.0	76.4	65.8
Norethindrone	93.4	93.2	96.3	103	95.7	94.5	106	93.2	102	89.1
Progesterone	85.3	83.1	150	107	71.7	63.4	92.6	92.3	63.2	60.7
Testosterone	99.8	99.3	95.8	100	121	96.7	102	96.1	99.1	92.9
trans-Diethylstilbestrol	99.6	97.4	94.2	95.1	91.9	88.4	7.8	97.9	94.9	90.2
					Isotope-dilution standards					
16-Epiestriol-d_2	60.0	70.5	48.3	33.6	55.5	76.6	69.4	76.4	70.5	71.7
17-*alpha*-Ethynylestradiol-d_4	65.8	72.5	79.0	96.1	54.1	76.7	79.2	75.1	81.0	71.8
17-*beta*-Estradiol-$^{13}C_6$	69.9	76.2	78.8	83.2	60.7	73.1	77.3	77.7	76.4	72.2
Bisphenol A-d_{16}	87.5	83.9	84.1	78.5	60.3	80.9	72.8	91.2	78.4	78.5
Cholesterol-d_7	78.5	82.1	71.6	80.9	85.2	72.6	66.0	57.2	62.3	64.3
Diethylstilbestrol-d_8	69.7	64.4	76.6	75.9	55.4	68.1	44.4	48.2	56.9	58.8
Estrone-$^{13}C_6$	70.6	79.6	86.7	92.7	56.8	80.2	71.8	78.5	80.6	70.9
Medroxyprogesterone-d_3	70.8	78.4	60.8	78.3	71.6	92.8	83.2	76.4	77.7	70.4
Mestranol-d_4	64.4	72.5	77.0	92.7	52.1	70.1	81.4	73.4	75.0	72.3
Nandrolone-d_3	77.1	84.9	105	113	50.1	75.1	84.6	84.9	81.0	73.3
Sample volume, milliliters	457	443	611	525	717	379	487	477	453	452

Table 29. Recoveries of method compounds in field-requested laboratory matrix-spike samples (FRLMS) and laboratory matrix-spike samples (MSPK) submitted in 2010 for analysis by laboratory schedules 2434 or 4434.—Continued

[RSD, relative standard deviation. All values in percent except as otherwise noted. Some values might have additional bias due to concentrations in the ambient sample between 10 and 100 percent (**bold** values) or 100 and 225 percent (***bold italicized*** values) of the fortified amount. Analyte fortification levels were 10 nanograms per liter (ng/L) for 17 analytes, 100 ng/L for bisphenol A, 320 or 1,000 ng/L for 3-*beta*-coprostanol, and 1,000 ng/L for cholesterol assuming a 0.5-L sample volume]

Analyte	Number of spike samples	Mean	Standard deviation	RSD	Median	F-pseudosigma	Relative F-pseudosigma	Minimum	Maximum
					Statistical summary for matrix-spike samples				
			Method analytes						
11-Ketotestosterone	23	77.8	32.0	41.2	91.3	19.5	21.3	11.5	123
17-*alpha*-Estradiol	23	101	7.1	7.0	102	8.0	7.9	86.4	111
17-*alpha*-Ethynylestradiol	23	91.7	6.2	6.8	92.1	6.8	7.3	79.8	100
17-*beta*-Estradiol	23	97.9	6.4	6.6	98.3	4.6	4.7	85.0	110
3-*beta*-Coprostanol	23	141	234	165	92.9	6.3	6.8	74.8	1,210
4-Androstene-3,17-dione	23	104	9.8	9.4	103	8.8	8.6	88.8	130
Bisphenol A	23	84.3	7.9	9.4	84.7	6.3	7.5	69.2	102
Cholesterol	23	121	143	119	92.6	9.7	10.5	65.2	776
cis-Androsterone	23	109	17.8	16.4	105	15.5	14.8	83.2	160
Dihydrotestosterone	23	100	14.1	14.0	96.1	8.3	8.6	78.5	134
Epitestosterone	23	107	8.9	8.3	105	8.4	7.9	94.8	129
Equilenin	23	80.9	16.8	20.7	83.5	14.8	17.7	41.2	107
Equilin	23	103	30.7	29.8	97.2	35.1	36.1	41.2	173
Estriol	23	71.5	30.7	42.9	82.3	15.9	19.3	6.1	105
Estrone	23	101	7.4	7.3	100	4.4	4.4	83.9	116
Mestranol	23	96.0	5.2	5.4	97.4	6.2	6.4	86.6	105
Norethindrone	23	93.2	5.2	5.5	93.3	1.9	2.1	81.8	106
Progesterone	23	66.0	59.3	89.9	70.1	45.3	64.6	0.0	271
Testosterone	23	103	9.0	8.7	99.9	7.2	7.2	94.5	131
trans-Diethylstilbestrol	23	85.3	18.7	21.9	90.5	8.3	9.2	7.9	99.6
			Isotope-dilution standards						
16-Epiestriol-d_2	23	65.8	13.1	19.9	65.9	10.8	16.5	33.6	91.4
17-*alpha*-Ethynyl estradiol-d_4	23	74.6	11.1	14.9	73.2	8.4	11.5	51.5	96.1
17-*beta*-Estradiol-$^{13}C_6$	23	75.1	8.9	11.9	75.3	8.5	11.2	57.7	89.3
Bisphenol A-d_{16}	23	81.6	12.9	15.8	84.1	10.0	11.9	54.3	104
Cholesterol-d_7	23	75.1	9.2	12.2	77.0	9.1	11.8	56.0	92.2
Diethylstilbesterol-d_8	23	53.5	15.1	28.3	54.2	16.2	29.9	19.9	76.6
Estrone-$^{13}C_6$	23	76.2	13.0	17.1	76.9	11.0	14.4	34.1	93.2
Medroxyprogesterone-d_3	23	62.2	24.1	38.7	70.8	18.8	26.6	9.2	92.8
Mestranol-d_4	23	73.5	11.0	15.0	72.5	9.0	12.4	51.6	92.7
Nandrolone-d_3	23	73.6	22.2	30.2	75.1	19.7	26.3	30.3	119
Sample volume, milliliters	23	520	146	28	460	46.7	10	375	931

[a] This reagent-water spike (RWS) or FRLMS sample was fortified with analytes only and stored frozen (−15 C) for 83 or 85 days before extraction. Isotope-dilution standards were added on day extracted. The RWS recoveries are not included in the statistical summary.

[b] Only this sample contained ascorbic acid

[c] Resample on July 29 of well water sampled on February 19 for the FRLMS 83 freeze-storage test matrix.

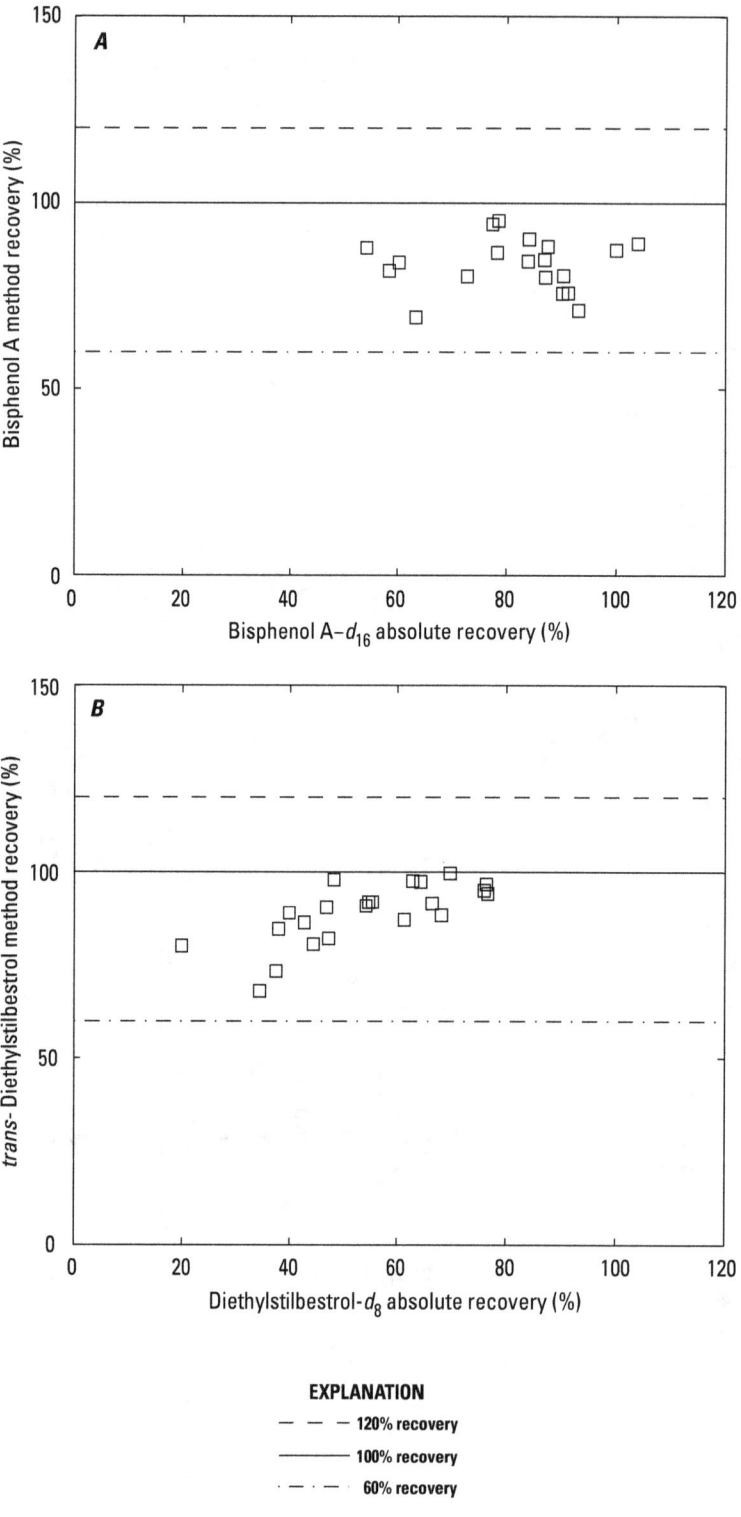

Figure 10. Relation between analyte method recoveries and isotope-dilution standard absolute recoveries in percent (%) in field-requested laboratory matrix-spike samples (FRLMS) or laboratory matrix-spike samples (MSPK) samples (see table 29). Samples with ambient analyte concentration that exceeded the fortified concentration were excluded to eliminate potential bias.

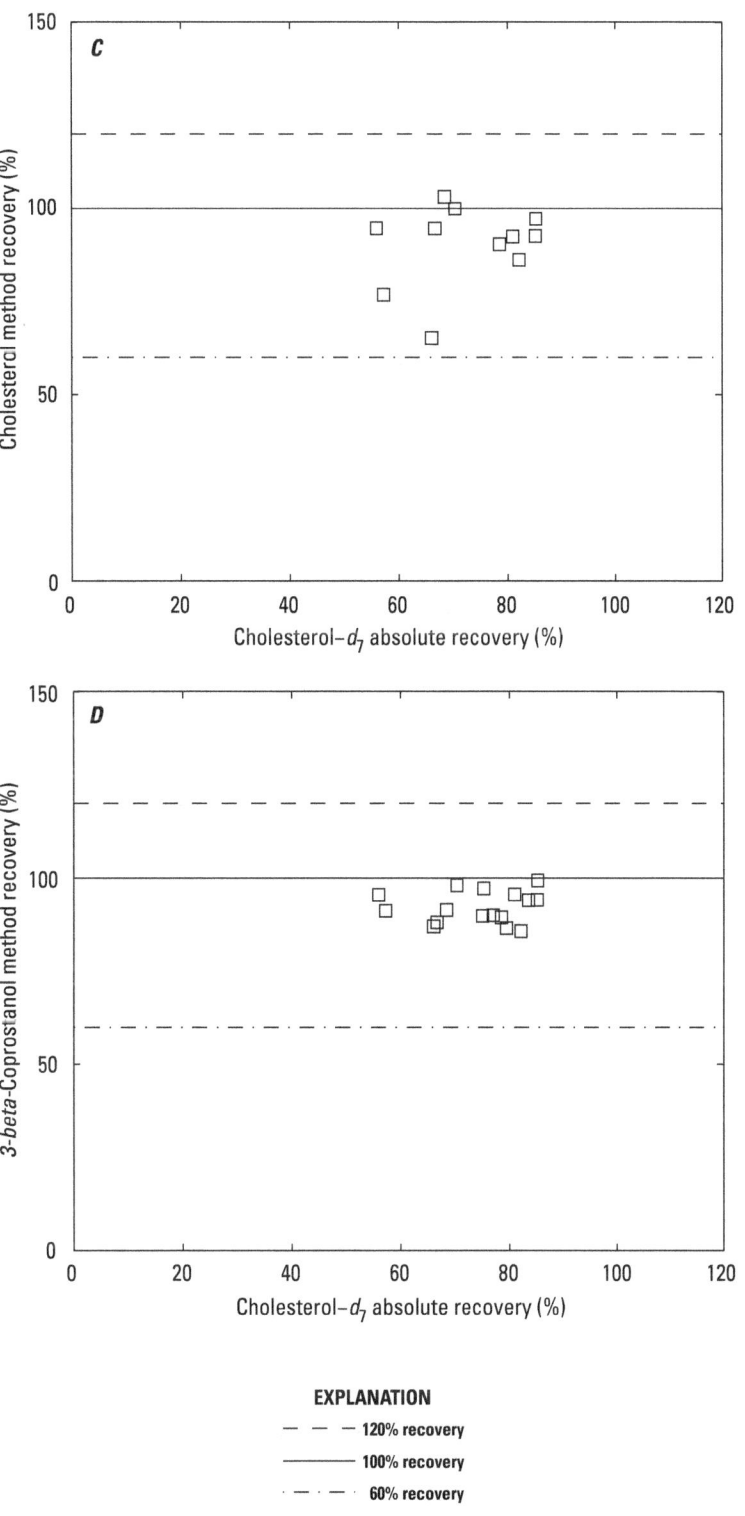

Figure 10. Relation between analyte method recoveries and isotope-dilution standard absolute recoveries in percent (%) in field-requested laboratory matrix-spike samples (FRLMS) or laboratory matrix-spike samples (MSPK) samples (see table 29). Samples with ambient analyte concentration that exceeded the fortified concentration were excluded to eliminate potential bias.—Continued

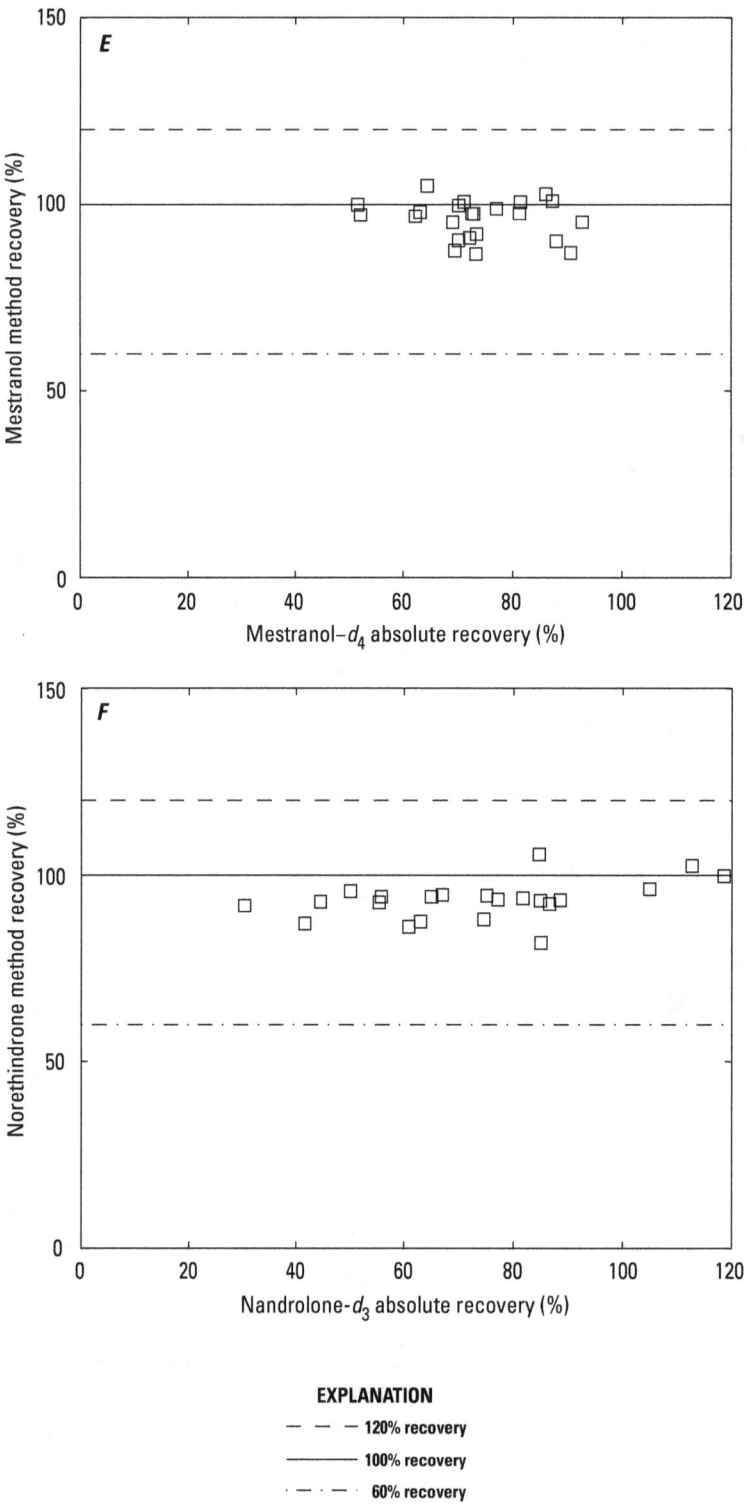

Figure 10.   Relation between analyte method recoveries and isotope-dilution standard absolute recoveries in percent (%) in field-requested laboratory matrix-spike samples (FRLMS) or laboratory matrix-spike samples (MSPK) samples (see table 29). Samples with ambient analyte concentration that exceeded the fortified concentration were excluded to eliminate potential bias.—Continued

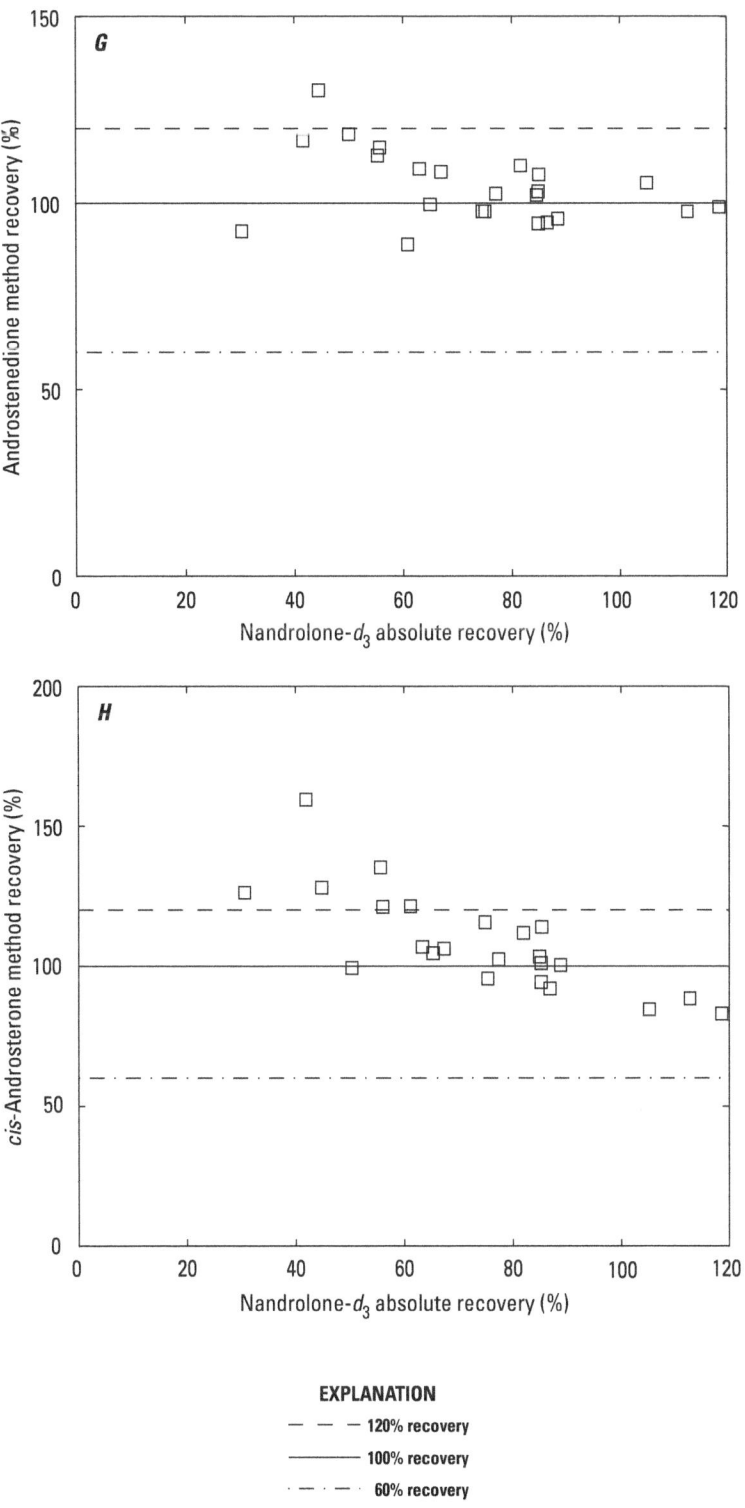

Figure 10. Relation between analyte method recoveries and isotope-dilution standard absolute recoveries in percent (%) in field-requested laboratory matrix-spike samples (FRLMS) or laboratory matrix-spike samples (MSPK) samples (see table 29). Samples with ambient analyte concentration that exceeded the fortified concentration were excluded to eliminate potential bias.—Continued

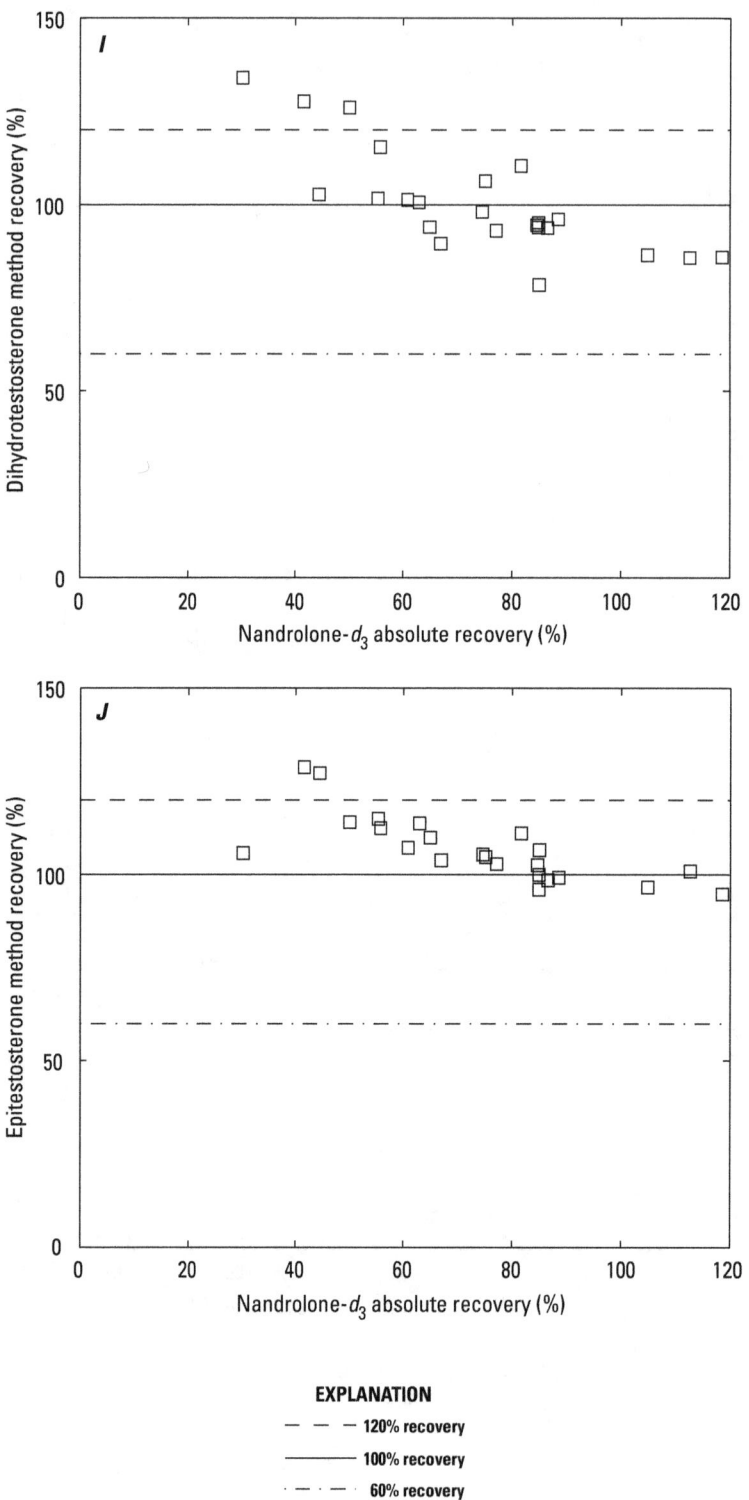

Figure 10. Relation between analyte method recoveries and isotope-dilution standard absolute recoveries in percent (%) in field-requested laboratory matrix-spike samples (FRLMS) or laboratory matrix-spike samples (MSPK) samples (see table 29). Samples with ambient analyte concentration that exceeded the fortified concentration were excluded to eliminate potential bias.—Continued

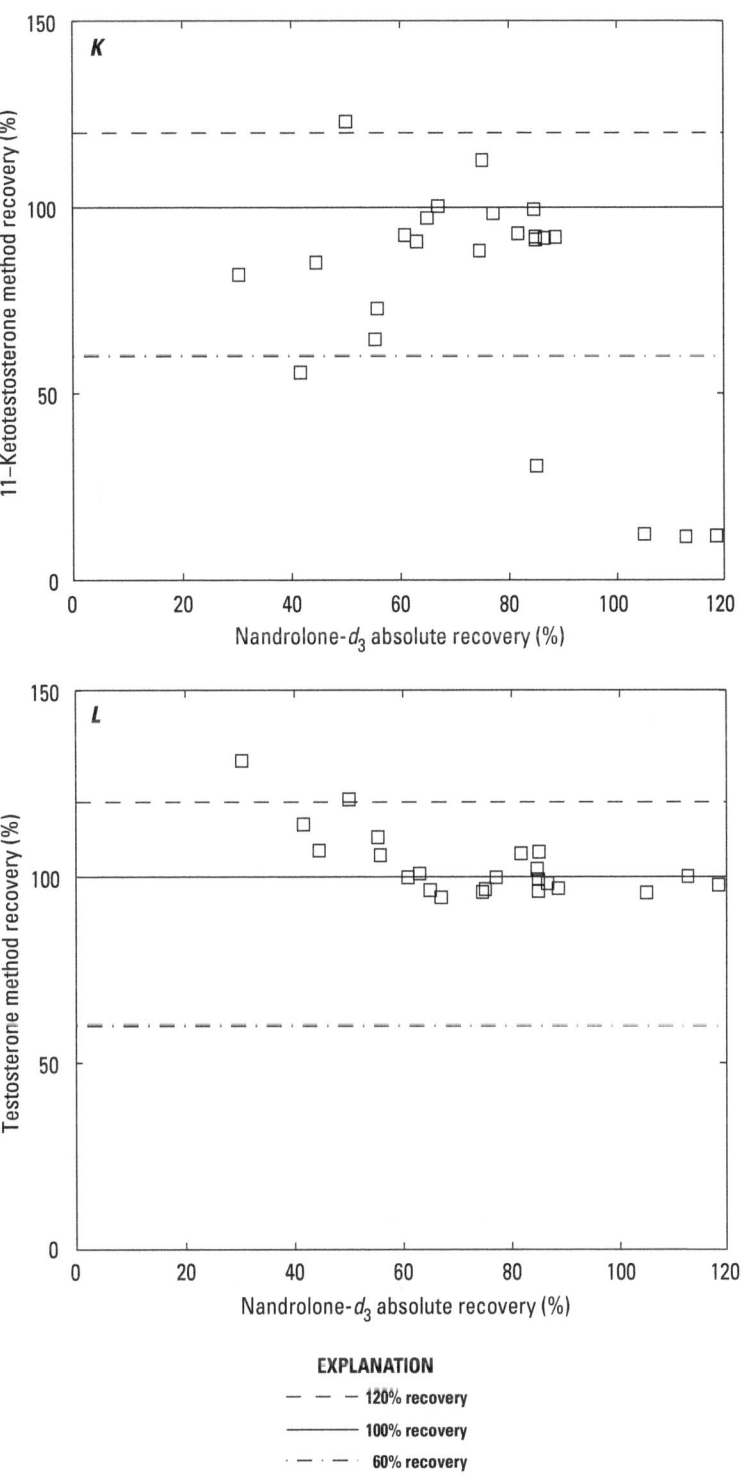

Figure 10. Relation between analyte method recoveries and isotope-dilution standard absolute recoveries in percent (%) in field-requested laboratory matrix-spike samples (FRLMS) or laboratory matrix-spike samples (MSPK) samples (see table 29). Samples with ambient analyte concentration that exceeded the fortified concentration were excluded to eliminate potential bias.—Continued

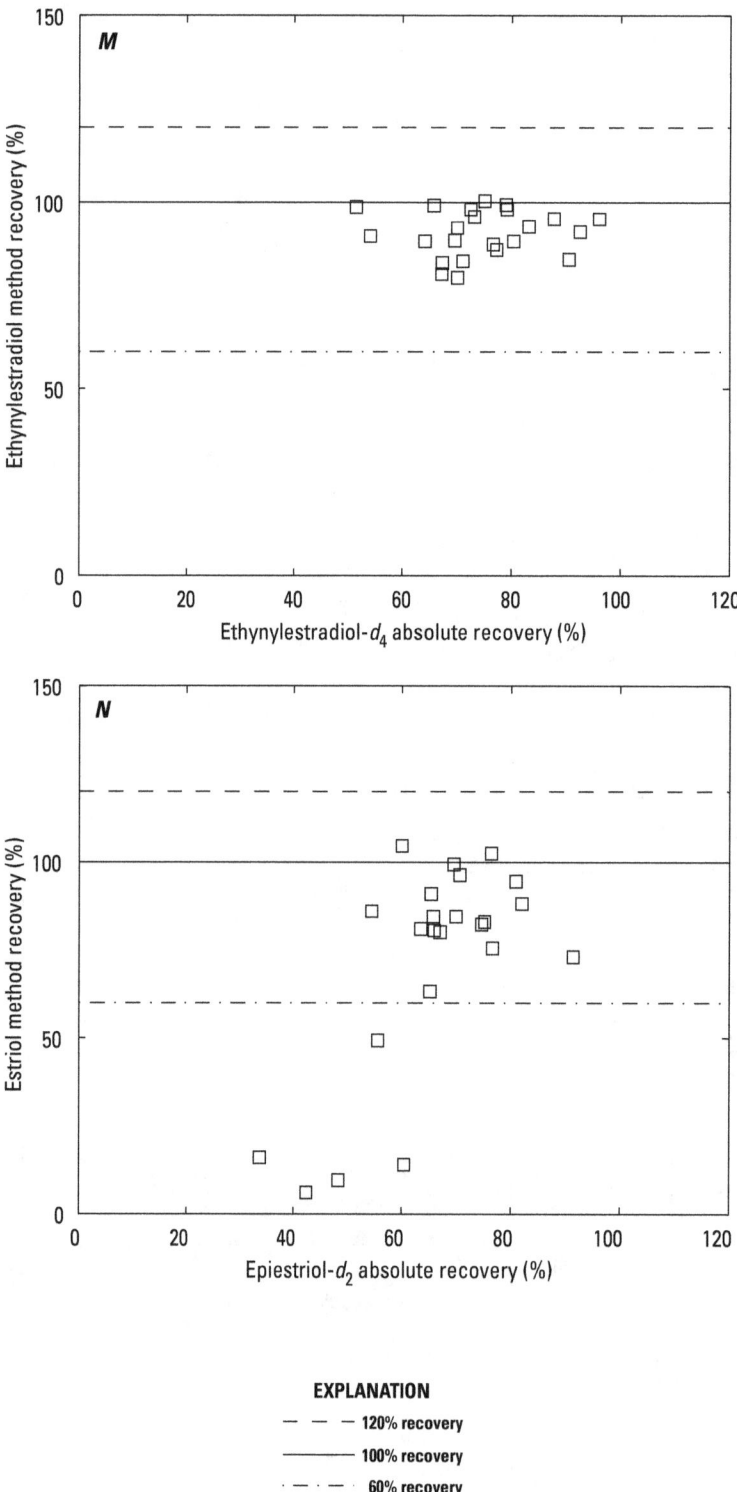

Figure 10.   Relation between analyte method recoveries and isotope-dilution standard absolute recoveries in percent (%) in field-requested laboratory matrix-spike samples (FRLMS) or laboratory matrix-spike samples (MSPK) samples (see table 29). Samples with ambient analyte concentration that exceeded the fortified concentration were excluded to eliminate potential bias.—Continued

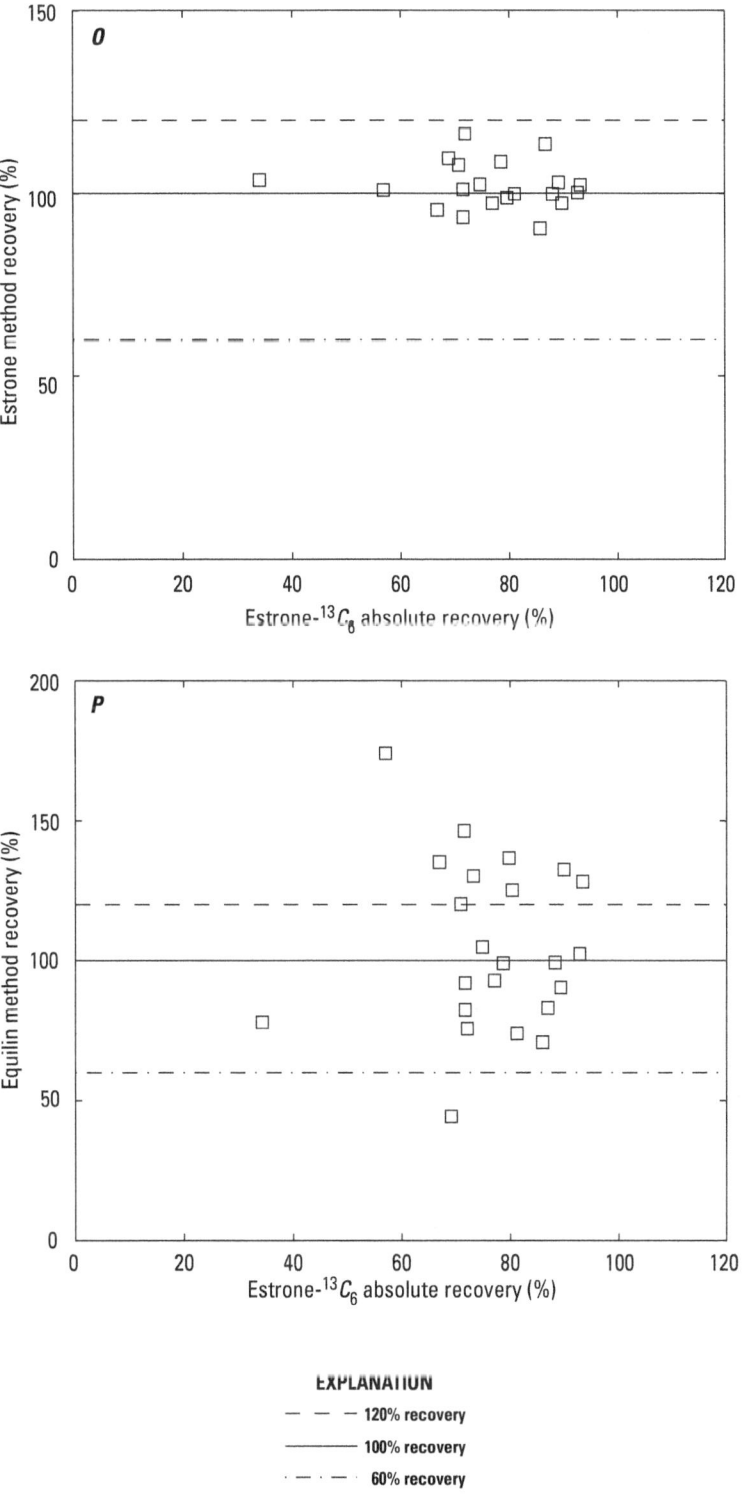

Figure 10.   Relation between analyte method recoveries and isotope-dilution standard absolute recoveries in percent (%) in field-requested laboratory matrix-spike samples (FRLMS) or laboratory matrix-spike samples (MSPK) samples (see table 29). Samples with ambient analyte concentration that exceeded the fortified concentration were excluded to eliminate potential bias.—Continued

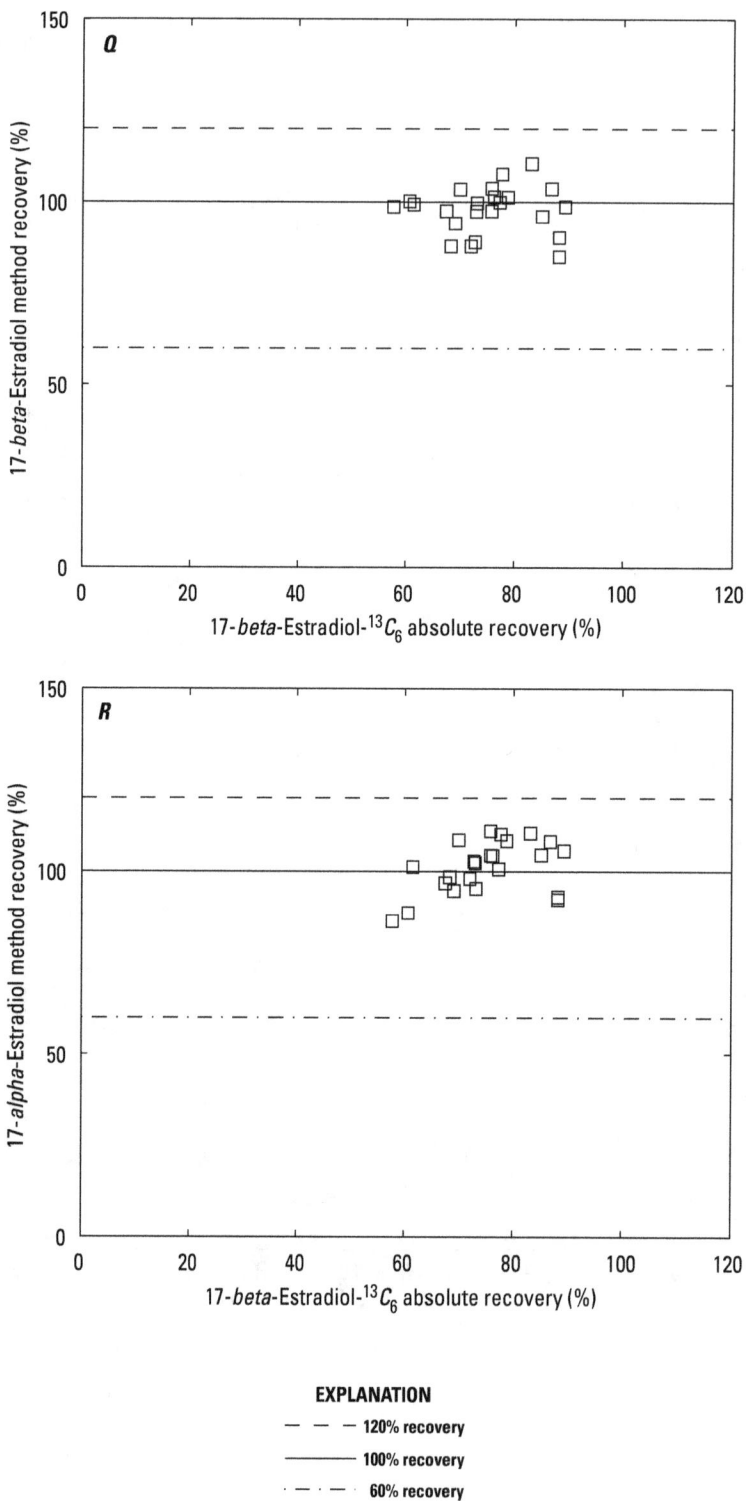

EXPLANATION

— — — 120% recovery
——— 100% recovery
— · — 60% recovery

Figure 10.   Relation between analyte method recoveries and isotope-dilution standard absolute recoveries in percent (%) in field-requested laboratory matrix-spike samples (FRLMS) or laboratory matrix-spike samples (MSPK) samples (see table 29). Samples with ambient analyte concentration that exceeded the fortified concentration were excluded to eliminate potential bias.—Continued

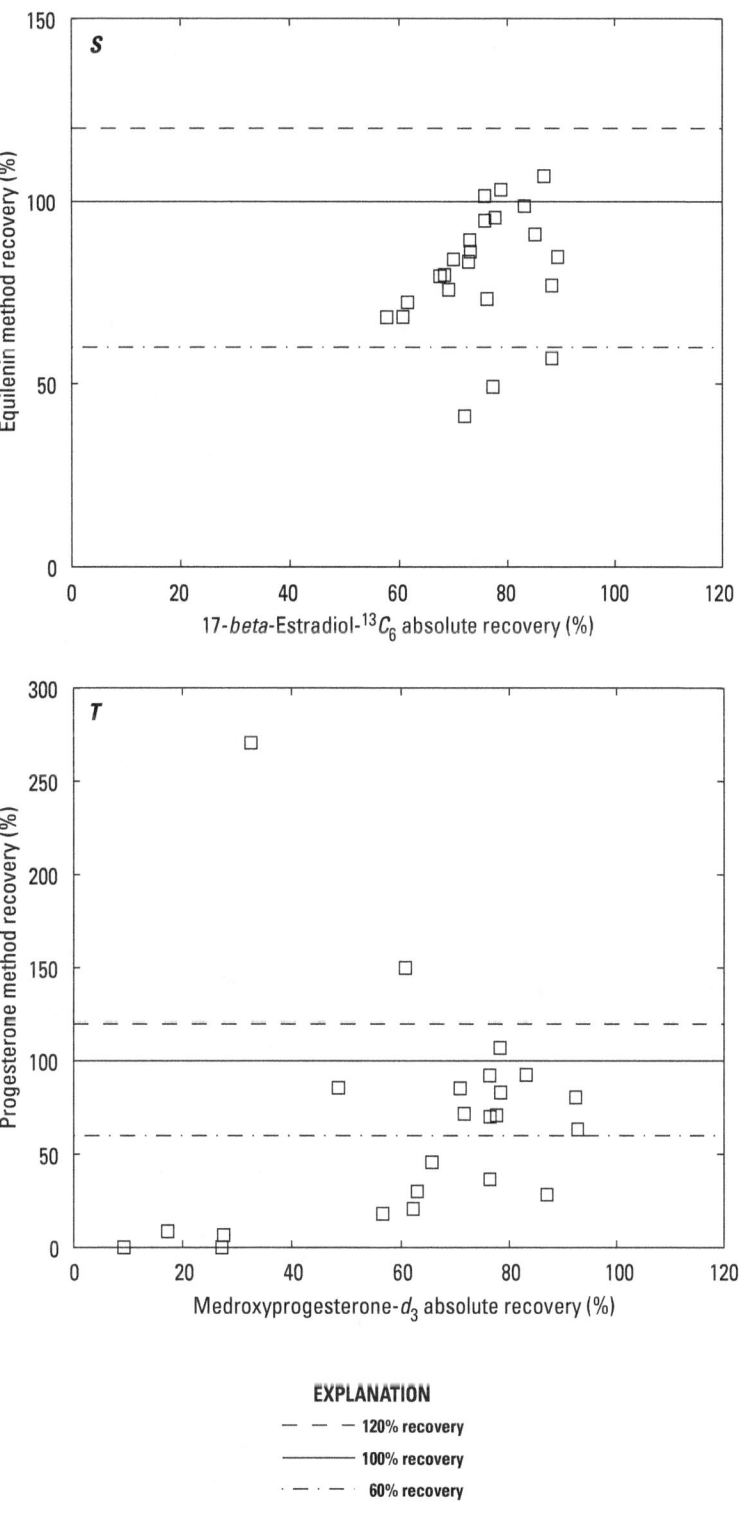

Figure 10. Relation between analyte method recoveries and isotope-dilution standard absolute recoveries in percent (%) in field-requested laboratory matrix-spike samples (FRLMS) or laboratory matrix-spike samples (MSPK) samples (see table 29). Samples with ambient analyte concentration that exceeded the fortified concentration were excluded to eliminate potential bias.—Continued

long as 85 days after spiking were well within the 60–120 percent recovery range and compare well to recoveries in the accompanied stored reagent-water spikes and to recoveries observed in the frozen-sample holding-time experiment (see "Method Analyte Stability in Frozen Reagent Water" section). Exceptions were recoveries of equilenin and, especially, *trans*-diethylstilbestrol in the stored groundwater spike FRLMS 83 (table 29). Equilenin's recovery in FRLMS 83 (49 percent) falls within the recovery range for the other matrix spikes (41–107 percent), but is lower than the other groundwater-only matrices (68–102 percent). Indeed, the collection well that was sampled for FRLMS 83 was resampled 5 months later on July 29, 2010, and an aliquot of this sample was fortified with the method analytes and IDS compounds just before extraction (refer to footnote c in table 29). The equilenin recovery was 68 percent in this July 29 resample.

Because equilenin does not have an exact IDS analog, it is unclear whether equilenin's absolute recovery during sample preparation and analysis was simply not well emulated by its IDS, estrone-$^{13}C_6$, in the FRLMS 83 matrix, or if it underwent some loss during the frozen-storage period. Conversely, *trans*-diethylstilbestrol clearly incurred loss during frozen storage of this groundwater sample, because its method recovery (8 percent) is dramatically less than the absolute recovery of its IDS diethylstilbestrol-d_8 (44 percent) added just before extraction. The method recovery of *trans*-diethylstilbestrol in the July 29, 2010, resample spike was 89 percent, whereas the absolute recovery of *trans*-diethylstilbestrol-d_8 was 38 percent and, although low, was similar to its recovery in FRLMS 83.

Although method recoveries for most analytes in the spiked matrices were within a range of 80–120 percent, the matrix-spike recovery results highlight that using an isotope-dilution quantification procedure still might be insufficient to compensate for matrix-specific performance limitations. More importantly, these results emphasize the importance of including matrix-spike samples as a quality-assurance component in environmental studies of these steroid hormones and related compounds.

Summary and Conclusions

The U.S. Geological Survey's National Water Quality Laboratory (NWQL) has developed a new analytical method for the determination of 20 steroid hormones and related compounds, many of which reportedly exhibit endocrine system modulating activity. The analytes include 6 natural and 3 synthetic estrogen compounds, 6 natural androgens, 1 natural and 1 synthetic progestin compound, and 2 sterols: cholesterol and 3-*beta*-coprostanol. These two sterols have limited biological activity but typically are abundant in wastewater effluents and serve as useful tracers. Bisphenol A, an industrial chemical used primarily to produce polycarbonate plastic and epoxy resins and that has been shown to have estrogenic activity, is also determined by the

method. The method is applicable to a variety of filtered or unfiltered water-matrix types including groundwater, surface water, surficial runoff, and wastewater-treatment plant (WWTP) effluent and influent samples. However, method performance for some analytes might be outside the desired recovery range of 60 to 120 percent, and some analytes have more variable performance (such as those described in this report) for some matrices including WWTP influents and primary effluents, biosolids runoff samples, animal-feeding operation waste lagoon samples, and other "complex" water samples.

Method analytes are determined in field-filtered or unfiltered water samples collected into 0.5-liter, high-density polyethylene bottles (containing ascorbic acid if suspected to be a chlorinated/brominated treated-water sample). Deuterium- or carbon-13-labeled isotope-dilution standards (IDSs), all of which are direct or chemically similar isotopic analogs of the method analytes, are added to the samples before analyte isolation by solid-phase extraction of the water sample using a octadecylsilyl (C_{18}) silica sorbent disk that is overlain with a graded glass-fiber filter to facilitate extraction of unfiltered sample matrices. Method analytes are eluted from the disk with methanol. The resultant extract is evaporated, reconstituted, and passed through a Florisil solid-phase extraction column to remove polar organic interferences in the extract. The resultant extract eluent is evaporated to dryness and the method compounds in the samples and in associated calibration standards are derivatized to trimethylsilyl analogs. These analogs are separated by gas chromatography and detected by tandem-quadrupole mass spectrometry by monitoring the product ions of three specific precursor-to-product ion transitions (two transitions for the IDSs). All 20 method analytes are quantified relative to a specific IDS compound by using an isotope-dilution quantification procedure.

Method performance was tested by spiking replicates of the following sample validation matrices at 10 and 100 nanograms per liter (ng/L) for most analytes: reagent water, wastewater-affected surface water, WWTP secondary effluent, and WWTP primary effluent (no biological treatment). For most analytes, mean method recoveries in these matrices were within the desired recovery performance range of 60–120 percent; relative standard deviations of recovery typically were no more than 25 percent. Exceptions occurred in the field matrices (particularly the primary WWTP effluent matrix) for those analytes that had substantial ambient concentrations relative to the analyte fortification level, which leads to enhanced recovery bias, variability, or both. Matrix-spike samples of additional field matrices provided similar results to those obtained for the validation matrices. Progesterone had unusually low recoveries in some matrices (especially some surface waters). Equilin had more variable recoveries in spiked matrices. Thus, determined sample concentrations of progesterone or equilin are reported as estimated to the National Water Information System (NWIS) using an "E" result-level remark code.

Bisphenol A, cholesterol, and 3-*beta*-coprostanol are sample preparation blank-limited analytes, and 11-ketotestosterone is an instrumental blank-limited analyte. Concentrations of these four analytes are reported using the minimum reporting level (MRL) convention; no concentrations are reported below the MRL concentration. The other 16 method analytes are reported using the laboratory reporting level convention with an interim reporting level (IRL) type code. Analytes meeting mass-spectral-identification criteria that have determined concentrations that are less than the reporting level, and even less than the applied detection level, are reported. Detection levels in reagent water for the 16 non-blank-limited analytes were determined using ASTM International's multi-concentration Interlaboratory Detection Estimate procedure and calculator, which estimates both the U.S. Environmental Protection Agency's method detection limit and Currie's critical level values. Based on these estimates, applied detection level values range from 0.4 to 4 ng/L for the 16 non-blank-limited analytes; applied reporting levels range from 0.8 to 8 ng/L.

The absolute recoveries of IDS compounds added to each sample before extraction also are reported to NWIS (along with determined analyte concentrations) to provide an indication of procedural performance for the specific sample in a manner comparable to surrogate compound recoveries provided by other NWQL methods. The IDS absolute recoveries will differ from, and normally be substantially less than, the determined method recoveries for analytes in spiked matrices because the analyte recovery is corrected during the isotope-dilution quantification procedure by use of the IDS's absolute recovery in the sample. Reported analyte concentrations in samples are automatically recovery-corrected by using isotope-dilution quantification. Qualification of reported analyte data in NWIS is based on sample-specific IDS-recovery information and performance criteria described in this report.

Several deuterium-labeled compounds initially tested as candidate direct-analog IDS compounds were determined to be unacceptable because either they did not have sufficient chemical purity or were susceptible to deuterium loss (deuterium-hydrogen exchange) in protic solvents, which compromises accuracy of the IDS correction. Careful consideration of label type, position, and stability, along with isotope purity, is vital when evaluating any labeled-compound as an IDS or surrogate candidate, especially for methods that quantify analytes at extremely low concentrations in a wide variety of matrices. Several analytes had no exact IDS analog available or, if available, were not tested because of excessive cost. Additional labeled compounds might be added as method IDS compounds as they become available in the future to further improve quantitative accuracy for those analytes described in this report that did not have exact isotopic analogs as of April 2012.

Holding-time experiments indicate overall acceptable analyte stability in reagent water stored refrigerated (4 degrees Celsius) for as long as 8 days and stored in a freezer

(–15 degrees Celsius) for as long as 56 days. Freezer storage of samples before extraction is used by the NWQL and encouraged for field storage to reduce microbiotic or other degradation processes. The 0.5-liter high-density polyethylene bottle is used as the sample container for this method to facilitate freeze storage of the samples prior to extraction.

Many of the method analytes are naturally occurring compounds, and bisphenol A is a component of polycarbonate plastic and epoxy resin materials used in a variety of products. As such, the inclusion of field blanks during sampling is vital to assess the potential for unintended contamination of samples with these analytes. Likewise, the matrix-spike results presented in this report highlight the importance of including field-submitted matrix-spike samples as a quality-assurance component in environmental studies that use this method.

References Cited

Ankley, G.T., Jensen, K.M., Makynen, E.A., Kahl, M.D., Korte, J.J., Hornung, M.W., Henry, T.R., Denny, J.S., Leino, R.L., Wilson, V.S., Cardon, M.C., Hartig, P.C., and Gray, L.E., 2003, Effects of the androgenic growth promoter 17-β-trenbolone on fecundity and reproductive endocrinology of the fathead minnow: Environmental Toxicology and Chemistry, v. 22, p. 1,350–1,360.

Antignac, J.P., Le Bizec, B., Monteau, F., and Andre, F., 2003, Validation of analytical methods based on mass spectrometric detection according to the "2002/657/EC" European decision—Guideline and application: Analytica Chimica Acta, v. 483, p. 325–334.

ASTM International, 2007, Standard practice for 99 %/95 % interlaboratory detection estimate (IDE) for analytical methods with negligible calibration error: West Conshohocken, Pa., ASTM International, ASTM D6091–07, 13 p., accessed online April 2012, at *http://www.astm.org/Standards/D6091.htm*.

ASTM International, 2008, Standard practice for the estimation of holding time for water samples containing organic and inorganic constituents: West Conshohocken, Pa., ASTM International, ASTM D4841–88 (reapproved 2008), 13 p., accessed online April 2012, at *http://www.astm.org/Standards/D4841.htm*.

ASTM International, 2010, Standard practice for performing detection and quantitation estimation and data assessment utilizing DQCALC software, based on ASTM practices D6091 and D6512 of Committee D19 on water: West Conshohocken, Pa., ASTM International, ASTM D7510–10, 2 p., accessed online April 2012, at *http://www.astm.org/Standards/D7510.htm*.

Ayebo, A., Breuer, G.M., Cain, T.G., Wichman, M.D., Subramanian, P., and Reynolds, S.J., 2006, Sterols as bio-markers for waste impact and source characterization in stream sediment: Journal of Environmental Health, v. 68, no. 10, p. 46–50.

Barber, L.B., Brown, G.K., and Zaugg, S.D., 2000, Potential endocrine disrupting organic chemicals in treated municipal wastewater and river water, chap. 7 *of* Keith, L.H., Jones-Lepp, T.L., and Needham, L.L., eds., Analysis of environmental endocrine disruptors: Washington, D.C., American Chemical Society Symposium Series 747, p. 97–123.

Behrman, E.J., and Gopalan, V., 2005, Cholesterol and plants: Journal of Chemical Education, v. 82, no. 12, p. 1791–1793.

Belchetz, P.E., 1994, Hormonal treatment of postmenopausal women: New England Journal of Medicine, v. 330, p. 1062–1071.

Caldwell, D.J., Mastrocco, F., Hutchinson, T.H., Lange, R., Heijerick, D., Janssen, C., Anderson, P.D., Sumpter, J.P., 2008, Derivation of an aquatic predicted no-effect concentration for the synthetic hormone 17α-ethinyl estradiol: Environmental Science and Technology, v. 42, p. 7046–7054.

Carpinteiro, J., Quintana, J.B., Rodriguez, I., Carro, A.M., Lorenzo, R.A., and Cela, R., 2004, Applicability of solid-phase microextraction followed by on-fiber silylation for the determination of estrogens in water samples by gas chromatography-tandem mass spectrometry: Journal of Chromatography A, v. 1056, p. 179–185.

Childress, C.J.O., Foreman, W.T., Connor, B.F., and Maloney, T.J., 1999, New reporting procedures based on long-term method detection levels and some considerations for interpretation of water-quality data provided by the U.S. Geological Survey National Water Quality Laboratory: U.S. Geological Survey Open-File Report 99–193, 19 p., accessed April 2012, at *http://water.usgs.gov/owq/OFR_99-193/index.html*.

De Llasera, M.P.G., Rodriguez-Castillo, A., and Vera-Avila, L.E., 2007, Relative influence of the dissolved humic material on the solid-phase extraction efficiency of pesticides from environmental water: Journal of Environmental Science and Health, part B—Pesticides Food Contaminants and Agricultural Wastes, v. 42, p. 615–627.

Fine, D.D., Breidenbach, G.P., Price, T.L., and Hutchins, S.R., 2003, Quantitation of estrogens in ground water and swine lagoon samples using solid-phase extraction, pentafluorobenzyl/trimethylsilyl derivatization and gas chromatography-negative ion chemical ionization tandem mass spectrometry: Journal of Chromatography A, v. 1017, p. 167–185.

Foreman, W.T., and Foster, G.D., 1991, Isolation of multiple classes of pesticides from large-volume water samples using solid phase extraction cartridges, *in* Mallard, G.E., and Aronson, D.A., eds., U.S. Geological Survey Toxic Substances Hydrology Program—Proceedings of the technical meeting, Monterey, California, March 11–15, 1991: U.S. Geological Survey Water-Resources Investigations Report 91–4034, p. 530–533.

Foreman, W.T., ReVello, R.C., and Gray, J.L., 2010, Deuterium exchange complicates isotope dilution methods for steroid hormones, *in* SETAC North America 31st Annual Meeting Abstract Book, Portland, Oreg., November 7–11, 2010: Pensacola, Fla., Society of Environmental Toxicology and Chemistry, abstract no. 670.

Furlong, E.T., Gray, J.L., Quanrud, D.M., Teske, S.S., Werner, S.L., Esposito, K., Marine, J., Ela, W.P., Zaugg, S.D., Phillips, P.J., and Stinson, B., 2011, Hormones, pharmaceuticals, anthropogenic waste indicators, and total estrogenicity in liquid and solid samples collected to estimate the fate of estrogenic compounds during municipal sludge stabilization and dewatering: U.S. Geological Survey Open-File Report 2011–1132, 77 p. (Also available at *http://pubs.usgs.gov/of/2010/1132/*.)

Havens, S.M., Hedman, C.J., Hemming, J.D.C., Mieritz, M.G., Shafer, M.M., and Schauer, J.J., 2010, Stability, preservation, and quantification of hormones and estrogenic and adrogenic activities in surface water runoff: Environmental Toxicology and Chemistry, v. 29, no. 11, p. 2481–2490.

Helsel, D.R., and R. M. Hirsch, 2002, Statistical methods in water resources: U.S. Geological Survey Techniques of Water-Resources Investigations, Book 4, Chapter A3, 522 p. (Also available at *http://pubs.usgs.gov/twri/twri4a3/#pdf.*)

Hoaglin, D.C., Mosteller, F., and Tukey, J.W., 1983, Understanding robust and exploratory data analysis: New York, John Wiley, 447 p.

Hohenblum, P., Gans, O., Moche, W., Scharf, S., and Lorbeer, G., 2004, Monitoring of selected estrogenic hormones and industrial chemicals in groundwaters and surface waters in Austria: Science of the Total Environment, v. 333, no. 1–3, p. 185–193.

Horii, Y., and Kannan, K., 2008, Survey of organosilicone compounds, including cyclic and linear siloxanes, in personal-care and household products: Archives of Environmental Contamination and Toxicology, v. 55, no. 4, p. 701–710.

Huang, C.H., and Sedlak, D.L., 2001, Analysis of estrogenic hormones in municipal wastewater effluent and surface water using enzyme-linked immunosorbent assay and gas chromatography/tandem mass spectrometry: Environmental Toxicology and Chemistry, v. 20, no. 1, p. 133–139.

Ingrand, V., Herry, G., Beausse, J., and de Roubin, M.R., 2003, Analysis of steroid hormones in effluents of wastewater treatment plants by liquid chromatography-tandem mass spectrometry: Journal of Chromatography A, v. 1020, p. 99–104.

Jobling, S., Nolan, M., Tyler, C.R., Brighty, G.C., and Sumpter, J.P., 1998, Widespread sexual disruption in wild fish: Environmental Science and Technology, v. 32, no. 17, p. 2498–2506.

Kelly, C., 2000, Analysis of steroids in environmental water samples using solid-phase extraction and ion-trap gas chromatography-mass spectrometry and gas chromatography-tandem mass spectrometry: Journal of Chromatography A, v. 872, no. 1–2, p. 309–314.

Kidd, K.A., Blanchfield, P.J., Mills, K.H., Palace, V.P., Evans, R.E., Lazorchak, J.M., and Flick, R.W., 2007, Collapse of a fish population after exposure to a synthetic estrogen: Proceedings of the National Academy of Sciences of the United States of America, v. 104, p. 8897–8901.

Kolodziej, E.P., Gray, J.L., and Sedlak, D.L., 2003, Quantification of steroid hormones with pheromonal properties in municipal wastewater effluent: Environmental Toxicology and Chemistry, v. 22, p. 2622–2629.

Kumar, V., Nakada, N., Yasojima, M., Yamashita, N., Johnson, A.C., and Tanaka, H., 2011, The arrival and discharge of conjugated estrogens from a range of different sewage treatment plants in the UK: Chemosphere, v. 82, no. 8, p. 1124–1128.

Landrum, P.F., and Giesy, J.P., 1981, Anomalous breathrough of benzo(a)pyrene during concentration with Amberlite XAD-4 resin from aqueous solutions, in Keith, L.H., ed., Advances in the identification and analysis of organic pollutants in water: Ann Arbor, Mich., Ann Arbor Science, p. 345–355.

Lange, R., Hutchinson, T.H., Croudace, C.P., Sigmund, F., Schweinfurth, H.A., Hampe, P., Panter, G.H., and Sumpter, J.P., 2001, Effects of the synthetic estrogen 17α-ethinylestradiol on the life-cycle of the fathead minnow (*Pimephales promelas*): Environmental Toxicology and Chemistry, v. 20, no. 6, p. 1216–1227.

Lee, K.E., Langer, S.K., Barber, L.B., Writer, J.H., Ferrey, M.L., Schoenfuss, H.L., Furlong, E.T., Foreman, W.T., Gray, J.L., ReVello, R.C., Martinovic, D., Woodruff, O.P., Keefe, S.H., Brown, G.K., Taylor, H.E., Ferrer, I., and Thurman, E.M., 2011, Endocrine active chemicals, pharmaceuticals, and other chemicals of concern in surface water, wastewater-treatment plant effluent, and bed sediment, and biological characteristics in selected streams, Minnesota—Design, methods, and data, 2009: U.S. Geological Survey Data Series 575, 54 p., with appendixes. (Also available at *http://pubs.usgs.gov/ds/575/*.)

Leeming, R., Ball, A., Ashbolt, N., and Nichols, P., 1996, Using faecal sterols from humans and animals to distinguish faecal pollution in receiving waters: Water Research, v. 30, no. 12, p. 2893–2900.

Liu, Z.H., Kanjo, Y., and Mizutani, S., 2009a, Urinary excretion rates of natural estrogens and androgens from humans, and their occurrence and fate in the environment—A review: Science of the Total Environment, v. 407, no. 18, p. 4975–4985.

Liu, Z.H., Kanjo, Y., and Mizutani, S., 2009b, Removal mechanisms for endocrine disrupting compounds (EDCs) in wastewater treatment–physical means, biodegradation, and chemical advanced oxidation: A review: Science of the Total Environment, v. 407, no. 2, p. 731–748.

Maloney, T.J., ed., 2005, Quality management system, U.S. Geological Survey National Water Quality Laboratory: U.S. Geological Survey Open-File Report 2005–1263, version 1.3, 9 November 2005, chapters and appendixes variously paged. (Also available at *http://pubs.usgs.gov/of/2005/1263/*.)

McGee, M.R., Julius, M.L., Vajda, A.M., Norris, D.O., Barber, L.B., and Schoenfuss, H.L., 2009, Predator avoidance performance of larval fathead minnows (*Pimephales promelas*) following short-term exposure to estrogen mixtures: Aquatic Toxicology, v. 91, no. 4, p. 355–361.

Merck Chemicals Company, 2010, Safety data sheet for 2,2,2-trifluoro-*N*-methyl-*N*-(trimethylsilyl)-acetamide (MSTFA): Darmstadt, Germany, Merck Chemicals Company, version 4.6, accessed online April 2012, at *http://www.merck-chemicals.com/2-2-2-trifluoro-n-methyl-n-trimethylsilylacetamide/MDA_CHEM-111805/p_uuid?attachments=MSDS*. [dated 5/11/2010]

Mills, L.J., and Chichester, C., 2005, Review of evidence—Are endocrine-disrupting chemicals in the aquatic environment impacting fish populations?: Science of the Total Environment, v. 343, no. 1–3, p. 1–34.

Mittendorf, R., 1995, Teratogen update—Carcinogenesis and teratogenesis associated with exposure to diethylstilbestrol (DES) in utero: Teratology, v. 51, no. 6, p. 435–445.

National Water Quality Laboratory, 2011, Requirements for the proper shipping of samples to the National Water Quality Laboratory: Denver, Colo., National Water Quality Laboratory Technical Memorandum 2011.01, accessed online April 2012, at *http://nwql.usgs.gov/Public/tech_memos/nwql.2011-01.pdf*.

Phillips, P.J., Smith, S.G., Kolpin, D.W., Zaugg, S.D., Buxton, H.T., and Furlong, E.T., 2010, Method description, quality assurance, environmental data, and other information for analysis of pharmaceuticals in wastewater-treatment-plant effluents, streamwater, and reservoirs, 2004–2009: U.S. Geological Survey Open-File Report 2010–1102, 36 p. (Also available at *http://pubs.usgs.gov/of/2010/1102/*.)

PlasticsEurope, 2007, Applications for bisphenol A: Brussels, Belgium, European Information Centre on Bisphenol A, 33 p., accessed online April 2012, at *http://bisphenolaeurope.com/uploads/BPA%20applications.pdf*.

Rajapakse, N., Silva, E., Scholze, M., and Kortenkamp, A., 2004, Deviation from additivity with estrogenic mixtures containing 4-nonylphenol and 4-*tert*-octylphenol detected in the E-SCREEN bioassay: Environmental Science and Technology, v. 38, no. 23, p. 6343–6352.

Regis Technologies, Inc., 2010, Material safety data sheet for *N*-Methyltrimethylsilyltrifluoroacetamide (MSTFA): Morton, Ill., Regis Technologies, Inc., MSDS number 270589d2, accessed online April 2012, at *http://www.registech.com/Library/MSDS/270589.pdf*. [dated 3/11/2010]

Rocke, D.M., and Lorenzato, S., 1995, A two-component model for measurement error in analytical chemistry: Technometrics, v. 37, p. 176–184.

Rodriguez-Mozaz, S., de Alda, M.J.L., and Barcelo, D., 2004, Picogram per liter level determination of estrogens in natural water and waterworks by a fully automated on-line solid-phase extraction-liquid chromatography-electrospray tandem mass spectometry method: Analytical Chemistry, v. 76, p. 6998–7006.

Routledge, E.J., Sheahan, D.A., Desbrow, C., Brighty, G.C., Sumpter, J.P., and Waldock, M.J., 1998, Identification of estrogenic chemicals in STW effluent—2. *in vivo* responses in trout and roach: Environmental Science and Technology, v. 32, no. 11, p. 1559–1565.

Sandstrom, M.W., 1995, Filtration of water-sediment samples for the determination of organic compounds: U.S. Geological Survey Water-Resources Investigations Report 95–4105, 13 p., accessed online April 2012, at *http://nwql.usgs.gov/Public/pubs/WRIR/WRIR-95-4105.pdf*.

Schenck, K., Williams, D., Dugan, N., Mash, H., Speth, T., Wymer, L., Rosenblum, L., Wiese, T., and Merriman, B., 2008, Evaluation of the removal of estrogens following chlorination, *in* SETAC North America 29th Annual Meeting Abstract Book, Tampa, Fla., November 16–20, 2008: Pensacola, Fla., Society of Environmental Toxicology and Chemistry, abstract no. WP115.

Sigma-Aldrich Company, 2010, N-Methyl-N-trimethylsilyltrifluoroacetamide activated II material safety data sheet, version 4.1: St. Louis, Miss., Sigma-Aldrich Company, accessed online August 2011, at *http://www.sigmaaldrich.com/catalog/ProductDetail.do?lang=en&N4=44156|FLUKA&N5=SEARCH_CONCAT_PNO|BRAND_KEY&F=SPEC*. [revised 8/24/2010]

Sumpter, J.P., and Johnson, A.C., 2005, Lessons from endocrine disruption and their application to other issues concerning trace organics in the aquatic environment: Environmental Science and Technology, v. 39, no. 12, p. 4321–4332.

Ternes, T.A., Stumpf, M., Mueller, J., Haberer, K., Wilken, R.-D., and Servos, M.R., 1999, Behavior and occurrence of estrogens in municipal sewage treatment plants—I. Investigations in Germany, Canada and Brazil: Science of the Total Environment, v. 228, p. 81–90.

Ternes, T.A., Andersen, H., Gilberg, D., and Bonerz, M., 2002, Determination of estrogens in sludge and sediments by liquid extraction and GC/MS/MS: Analytical Chemistry, v. 74, no. 14, p. 3498–3504.

Thorpe, K.L., Gross-Sorokin, M.Y., Johnson, I.R., Brighty, G.C., and Tyler, C.R., 2006, An assessment of the model of concentration addition for predicting the estrogenic activity of chemical mixtures in wastewater treatment works effluents: Environmental Health Perspectives, v. 114, supplement 1, p. 1142–1149.

Tilton, F., Benson, W.H., and Schlenk, D.K., 2002, Evaluation of estrogenic activity from a municipal wastewater treatment plant with predominantly domestic input: Aquatic Toxicology, v. 61, no. 3–4, p. 211–224.

Timme, P.J., 1995, National Water Quality Laboratory 1995 services catalog: U.S. Geological Survey Open-File Report 95–352, 120 p.

Tolgyesi, A., Verebey, Z., Sharma, V.K., Kovacsics, L., and Fekete, J., 2010, Simultaneous determination of corticosteroids, androgens, and progesterone in river water by liquid chromatography-tandem mass spectrometry: Chemosphere, v. 78, no. 8, p. 972–979.

U.S. Environmental Protection Agency, 1986, Guidelines establishing test procedures for the analysis of pollutants— Appendix B to Part 136—Definition and procedures for the determination of the method detection limit—Revision 1.11: U.S. Code of Federal Regulations, Title 40, accessed online November 2012, at *http://www.gpo.gov/fdsys/pkg/CFR-2012-title40-vol24/pdf/CFR-2012-title40-vol24-part136-appB.pdf*.

U.S. Environmental Protection Agency, 2007a, Method 1698—Steroids and hormones in water, soil, sediment, and biosolids by high-resolution gas chromatography/high-resolution mass spectrometry (HRGC/HRMS): Washington D.C., U.S. Environmental Protection Agency, EPA–821–R–08–003, 64 p., accessed online April 2012, at *http://water.epa.gov/scitech/methods/cwa/bioindicators/upload/2008_01_03_methods_method_1698.pdf.*

U.S. Environmental Protection Agency, 2007b, Method 8290A—Polychlorinated dibenzodioxins (PCDDs) and polychlorinated dibenzofurans (PCDFs) by high-resolution gas chromatography/high-resolution mass spectrometry (HRGC/HRMS): Washington D.C., U.S. Environmental Protection Agency, accessed online April 2012, at *http://www.epa.gov/epawaste/hazard/testmethods/sw846/pdfs/8290a.pdf.*

U.S. Environmental Protection Agency, 2010a, Method 539—Determination of hormones in drinking water by solid phase extraction (SPE) and liquid chromatography electrospray ionization tandem mass spectrometry (LC-ESI-MS/MS): Washington D.C., U.S. Environmental Protection Agency Office of Water, EPA Document No. 815–B–10–001, 36 p., accessed April 2012, at *http://water.epa.gov/scitech/drinkingwater/labcert/upload/met539.pdf.*

U.S. Environmental Protection Agency, 2010b, Stability of pharmaceutical, personal care products, steroids, and hormones in aqueous samples, POTW effluents, and biosolids: Washington D.C., U.S. Environmental Protection Agency Office of Water, EPA–820–R–10–008, 38 p., accessed online April 2012, at *http://water.epa.gov/scitech/methods/cwa/upload/methodsppcp.pdf.*

U.S. Environmental Protection Agency, 2011, Revisions to the Unregulated Contaminant Monitoring Regulation (UCMR 3) for public water systems: U.S. Federal Register, v. 76, no. 42, p. 11713–11737, accessed online April 2012, at *http://edocket.access.gpo.gov/2011/pdf/2011-4641.pdf.*

U.S. Geological Survey, 2010, Changes to the reporting convention and to data qualification approaches for selected analyte results reported by the National Water Quality Laboratory (NWQL): Office of Water Quality Technical Memorandum 2010.07, unpaged, accessed online April 2012, at *http://water.usgs.gov/admin/memo/QW/qw10.07.html.*

U.S. Geological Survey, 2011a, User's manual for the National Water Information System of the U.S. Geological Survey—Water-quality system, version 4.11, Appendix A.—Codes used in water-quality processing system: U.S. Geological Survey, accessed online April 2012, at *http://nwis.usgs.gov/currentdocs/qw/QW.user.book.html.* [last modified March 1, 2011].

U.S. Geological Survey, 2011b, Requirements for the proper shipping of samples to the National Water Quality Laboratory: U.S. Geological Survey, National Water Quality Technical Memorandum 2011.01, accessed online April 2012, at *http://nwql.usgs.gov/Public/tech_memos/nwql.2011-01.pdf.*

U.S. Geological Survey, 2011c, Application of the result-level 'v' value qualifier code and 'E' remark code to selected organic results reported by the National Water Quality Laboratory (NWQL): Office of Water Quality Technical Memorandum 2012.01, accessed online April 2012, at *http://water.usgs.gov/admin/memo/QW/qw12.01.pdf.*

Vajda, A.M., Barber, L.B., Gray, J.L., Lopez, E.M., Woodling, J.D., and Norris, D.O., 2008, Reproductive disruption in fish downstream from an estrogenic wastewater effluent: Environmental Science and Technology, v. 42, p. 3407–3414.

Vajda, A.M., Barber, L.B., Gray, J.L., Lopez, E.M., Bolden, A.M., Schoenfuss, H.L., and Norris, D.O., 2011, Demasculinization of male fish by wastewater treatment plant effluent: Aquatic Toxicology, v. 103, no. 3–4, p. 213–221.

Vandenberg, L.N., Maffini, M.V., Sonnenschein, C., Rubin, B.S., and Soto, A.M., 2009, Bisphenol-A and the great divide—A review of controversies in the field of endocrine disruption: Endocrine Reviews, v. 30, p. 75–95.

Vanderford, B.J., Mawhinney, D.B., Trenholm, R.A., Zeigler-Holady, J.C., and Snyder, S.A., 2011, Assessment of sample preservation techniques for pharmaceuticals, personal care products, and steroids in surface and drinking water: Analytical and Bioanalytical Chemistry, v. 399, no. 6, p. 2227–2234.

Watabe, Y., Kondo, T., Imai, H., Morita, M., Tanaka, N., and Hosoya, K., 2004, Reducing bisphenol A contamination from analytical procedures to determine ultralow levels in environmental samples using automated HPLC microanalysis: Analytical Chemistry, v. 76, p. 105–109.

Weschler, C.J., Langer, S., Fischer, A., Bekö, G., Toftum, J., and Clausen, G., 2011, Squalene and cholesterol in dust from Danish homes and daycare centers: Environmental Science and Technology, v. 45, p. 3872–3879.

Wilde, F.D., Radtke, D.B., Gibs, Jacob, and Iwatsubo, R.T., eds., 2004 with updates through 2009, Processing of water samples (ver. 2.2): U.S. Geological Survey, Techniques of Water-Resources Investigations, book 9, chap. A5, April 2004. (Also available at *http://pubs.water.usgs.gov/twri9A5/.*)

Wise, A., O'Brien, K., and Woodruff, T., 2011, Are oral contraceptives a significant contributor to the estrogenicity of drinking water?: Environmental Science and Technology, v. 45, p. 51–60.

Zeilenger, J., Steger-Hartman, T., Maser, E., Goller, S., Vonk, R., and Lange, R., 2009, Effects of synthetic gestagens on fish reproduction: Environmental Toxicology and Chemistry, v. 28, p. 2663–2670.